Renewable Energy; a Status Quo

RIVER PUBLISHERS SERIES IN RENEWABLE ENERGY

Volume 1

Series Editor

ERIC JOHNSON
Atlantic Consulting
Switzerland

The "River Publishers Series in Renewable Energy" is a series of comprehensive academic and professional books which focus on theory and applications in renewable energy and sustainable energy solutions. The series will serve as a multi-disciplinary resource linking renewable energy with society. The book series fulfils the rapidly growing worldwide interest in energy solutions. It covers all fields of renewable energy and their possible applications will be addressed not only from a technical point of view, but also from economic, social, political, and financial aspect.

Books published in the series include research monographs, edited volumes, handbooks and textbooks. The books provide professionals, researchers, educators, and advanced students in the field with an invaluable insight into the latest research and developments.

Topics covered in the series include, but are by no means restricted to the following:

- Renewable energy
- Energy Solutions
- Energy storage
- Sustainability
- Green technology

For a list of other books in this series, visit www.riverpublishers.com

Renewable Energy; a Status Quo

Ali Sayigh

Director General
World Renewable Energy Network (WREN), UK

Published 2016 by River Publishers
River Publishers
Alsbjergvej 10, 9260 Gistrup, Denmark
www.riverpublishers.com

Distributed exclusively by Routledge
4 Park Square, Milton Park, Abingdon, Oxon OX14 4RN
605 Third Avenue, New York, NY 10017, USA

First issued in paperback 2023

Renewable Energy; a Status Quo / by Ali Sayigh.

Routledge is an imprint of the Taylor & Francis Group, an informa business

Publisher's Note
The publisher has gone to great lengths to ensure the quality of this reprint but points out that some imperfections in the original copies may be apparent.

ISBN 13: 978-87-7022-940-1 (pbk)
ISBN 13: 978-87-93379-50-3 (hbk)
ISBN 13: 978-1-003-33926-7 (ebk)

Contents

Preface

The World Renewable Energy Network (WREN) and the World Renewable Energy Congress (WREC) has always published an annual update regarding the progress in renewable energy since 2000.

The book in the past used to be in the form of a magazine. Last year, 2015, it has been our pleasure to team with River Publishers. WREN has teamed with River Publishers to change the report into a book covering the year's events, WREN/WREC activities, and some important papers around the world selected for their value to our readers.

Chapters include: *The Urgent Necessity to Redouble Renewable Energy Output* by Ali Sayigh, UK; *Future Concepts in Solar Thermal Electricity Generation* by Dr M Roger from Almeria, Spain; *Reinventing Geothermal Energy* by Bodo von During, von During Group, Switzerland; *From Waste to Energy: Development and Use of Renewable Energy in Sewage Treatment Facilities in Hong Kong*, by DAO Kwan-ming, Keith from Hong Kong; *Cacao and Coffee Roasting using Concentrated Solar Energy: A Technology for Small Farms in Rural Areas of Peru*, by François Veynandt, Juan Pablo Perez Panduro, Miguel Hadzich, Concentrating Solar Technologies, and Building's Energy Efficiency, Ecole des Mines d'Albi, France; *Socioeconomic, Environmental and Social Impacts of a Concentrated Solar Power Energy Project in Northern Chile*, Rodriguez Serrano from Spain.

At the end of the book, WREN activies for the period 2014 and 2015 are briefly discussed and the Calendar of events for 2016 and 2017 are included.

Ali Sayigh

List of Contributors

Natalia Caldés *Energy Department, Energy Systems Analysis Unit, CIEMAT, Av. Complutense 40, E-28040 Madrid, Spain*

Kwan-ming (Keith) Dao *Senior Electrical & Mech. Engineer, Project Division, Drainage Services Department, The Government of the Hong Kong, Special Administrative Region 44/F, Revenue Tower, 5 Gloucester Road, Wan Chai, Hong Kong, China*

Cristina De La Rúa *Energy Department, Energy Systems Analysis Unit, CIEMAT, Av. Complutense 40, E-28040 Madrid, Spain*

Alberto Garrido *CEIGRAM, Research Centre for the Management of Agricultural and Environmental Risks, Av. Complutense s/n, E-28040 Madrid, Spain*

Miguel Hadzich *Concentrating Solar Technologies and Building's Energy Efficiency, Ecole des Mines d'Albi Campus Jarland, 81013 ALBI Cedex 9, France*

Yolanda Lechón *Energy Department, Energy Systems Analysis Unit, CIEMAT, Av. Complutense 40, E-28040 Madrid, Spain*

Juan Pablo Perez panduro *Concentrating Solar Technologies and Building's Energy Efficiency, Ecole des Mines d'Albi Campus Jarland, 81013 ALBI Cedex 9, France*

Irene Rodríguez *Energy Department, Energy Systems Analysis Unit, CIEMAT, Av. Complutense 40, E-28040 Madrid, Spain*

Marc Röger *DLR German Aerospace Center, Institute of Solar Research, Plataforma Solar de Almería, 04200 Tabernas, Spain*

Ali Sayigh *Chairman of WREC & Director General WREN, Chairman IEI, Flat 3, House 5-6, Clarendon Terrace, P O Box 362, Brighton BN2 1YH, UK*

François Veynandt *Concentrating Solar Technologies and Building's Energy Efficiency, Ecole des Mines d'Albi Campus Jarland, 81013 ALBI Cedex 9, France*

Bodo von Düring *CLEAG Geothermal, Lidostrasse 6, CH-6006, Luzern, Switzerland*

List of Figures

List of Tables

1

The Urgent Necessity to Redouble Renewable Energy Output

Ali Sayigh

Chairman of WREC & Director General WREN
Chairman IEI
Flat 3, House 5-6, Clarendon Terrace, P O Box 362,
Brighton BN2 1YH, UK

Abstract

It is evident to all that climate change *is* happening – the results can be seen in many countries. Floods, freaks storms, wind speeds of more than 80 mph, heat waves, droughts, rising sea levels, and disappearing glaciers are largely due to excessive use of fossil fuels.

Climate change acceleration began slowly in the 1970s but has now increased beyond our ability to stop it or reduce its impact. Using renewable energy effectively on a large scale will put an end or considerably slow down the changes in many parts of the world. This paper shows that some countries are making greater efforts than others. Installations of renewable energy systems in the year 1970s and 1980s were limited to kilowatts, while in the 2010s we speak in terms of megawatts. The cost of most renewable energy systems have been reduced considerably that they have reached parity with fossil fuels or are even cheaper. The most effective progress has been made in photovoltaic systems: the cost of turnkey installations say for 5 MW is $6 million. Governments in European countries are using Feed-in-Tariffs which has made the payback period of installing a large system less than 1.25 years. Similarly, Concentrated Solar Power, biomass, wind energy, and hydro-power have greatly improved payback periods. Countries such as Morocco have pledged to produce 40% of their electricity from renewable energy by 2020, while Austria has declared that by 2050 all its energy will come from renewable sources.

It is clear from the media and UN Reports that there is no country which is not utilizing renewable energy to some extent, but what is urgently needed is for this use to be redoubled immediately to prevent the earth heating by more than 2°.

While much is hoped from the outcome of the December 2015 Paris climate summit. Realistically, in the past, very few nations honored their pledges. A great deal of aid has been given to poor countries which are suffering from climate change; however, the donor nations have failed to restrict their own carbon emissions. Many poor countries feel they are being expected to forgo the industrial benefits which came from the industrial revolution powered by fossil fuels. This paper will outline the achievements so far renewable energy have made by the end of 2015.

1.1 Climate Change

It is beyond doubt that now climate change is happening and the global temperature is rising to the critical limit of 1.5°C, which is causing changes all over the world. A recent example in the UK is the recurrent large-scale flooding which has occurred in the winter of 2015/16 with an estimated costing of £1.3 billion, while over the last 10 years the total spent has been more than £3 billion. It is generally held that one of the major causes of climate change is the excessive emission of CO_2 into the atmosphere from transport, industry, and agriculture. From the figure below it can be seen that more than 40t/p (tonne per person) is now being emitted per annum. To keep the temperature rise below the critical limit the emissions must be reduced to at least half the current level (Figure 1.1).

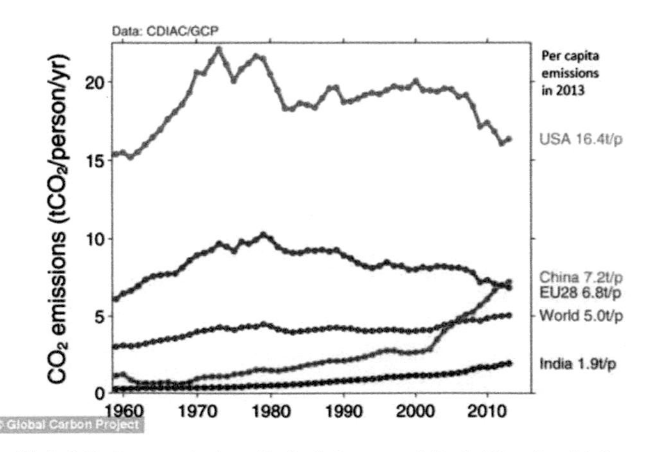

Figure 1.1 Global Carbon emission. Ref.: Science and Tech Monday 11 January 2016.

Figure 1.2 Coal power station in India.

How this can be achieved when coal is still being used extensively to generate electricity in China, India, and USA, despite the fact it causes 75% CO_2 emission which is a political and not a technological problem since solutions are available (Figure 1.2).

1.2 Renewable Energy Update

In this section, a full review of renewable energy is given showing the advances made to date and their low cost as well as their viability to supply all the power needed. For example in the late 1960s in Saudi Arabia, a 1 W **Photovoltaic** (PV) cell was impossible to obtain and by 1970s it would cost more than US$ 100 and by 1980 they were easily obtainable at a cost of US $ 10 per Watt.

In the year 1990s and 2000s, applications were specified in kW, now in 2015 the market talks about applications in MW only, see the example below from PHOTON Newsletter November 2015 (Figure 1.3).

Similarly, the case with wind energy, the most efficient and largest wind turbine in 1988 was 250 kW, while now 3 and 6 MW are common. Both PV and wind turbines efficiency has greatly improved, for example in the efficiency of crystalline PV panels has risen from 7 to 18%. Wind machines always had gear boxes which were costly and inefficient, nowadays most turbines are gearless. Similarly their cost used to be US$ 5000 perk W, now it is US$ 2000.

1.2.1 Wind Energy

Prior to 1980 most wind turbines were installed singly or in small groups with the aim of supplying electricity to small specific applications. Since 2005, however, most installations are in wind farms locations with up to dozens of turbines. They are now much less noisy, which meets one of the major

United PV aims to raise $100 million **ET Solar to build two PV plants totalling 10 MW in the UK** **Taiwan allows 500 MW FIT in 2016**

TSK and Enviromena to build 120 MW solar plant in Jordan Concord Green Energy **Spain's PV capacity reaches 4,667 MW** **Canadian solar park to sale of 9 MW**

Baywa r.e. sells 24.2 MW solar park Wirsol completes 61.2 MW India allocate 6.5 GW of PV power in the UK solar park in Denmark over the next three months through bond offering

Figure 1.3 PHOTON Newsletter 2015.

Figure 1.4 In 2014, GE supplied 235 turbines model GE 1.7–103 and related services for the 400 MW Texas Wind Project.

popular criticisms, more efficient and gearless. In term of sizes, many of them are between 2 and 3 MW while 6 and 7 MW capacity is available (Figure 1.4).

In 2015, Siemens Energy completed the largest wind farm in the world, more than 460 2.25 MW machines supplying 1,050-MW for the MidAmerican Energy Company (Figure 1.5).

Figure 1.5

Figure 1.6

Vestas in 2013 installed a 117-MW Tafila wind power plant 180 km south of Amman, Jordan, 38 V112-3.0-MW machines. Figure 1.7 shows 18,111-MW machines completed in 2014 in Estonia by GE. Figure 1.8 shows 676-MW wind turbines a total capacity of 402 MW located offshore in the East Coast of England. It is worth mentioning here that the UK has the largest Offshore wind farms in the world with a total capacity of 11,136 MW in 2014. Figure 1.9 shows the world's largest operating offshore wind project the London Array which produced a record 369 GWh in December, 2015 and an average of 78.9% capacity factor. (Ref.: Wind Power Offshore Weekly, 11 January 2016)

Figure 1.7

Figure 1.8

Figure 1.9

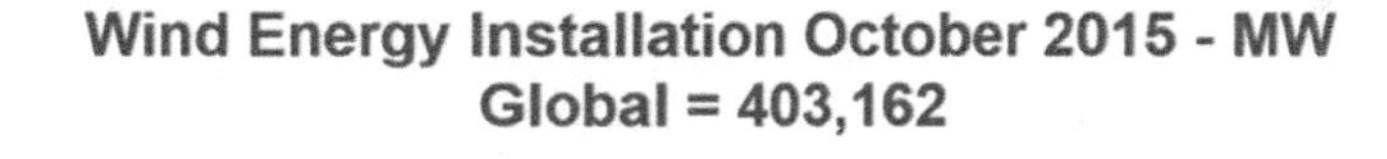

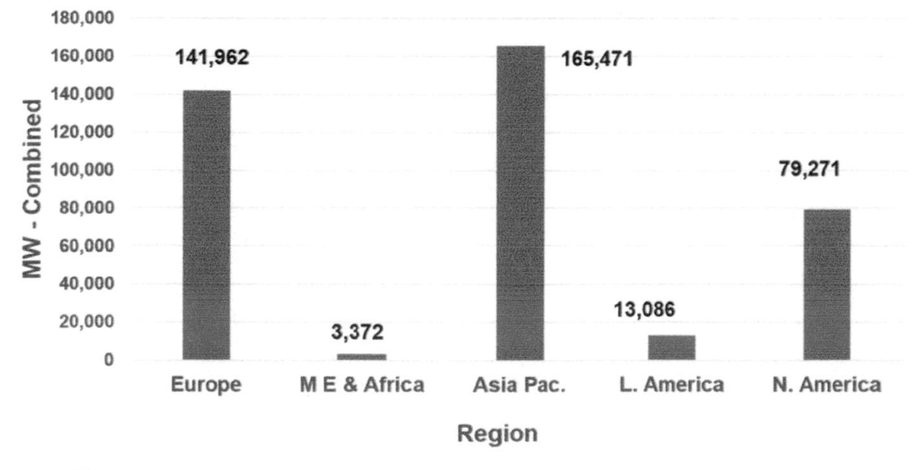

Figure 1.10 The Global installed Wind Power in October 2015 is 403, 162 MW.

While China remains the largest user of wind energy as shown in the figure below October 2015, it meets only a fraction of their electricity requirements, Denmark on the other hand in 2015 produced 42% of its electricity from wind energy as reported recently in the Guardian newspaper, indeed it was also reported that on one day in September Denmark did not use any of its central power stations and it is able to export some of their electricity abroad.

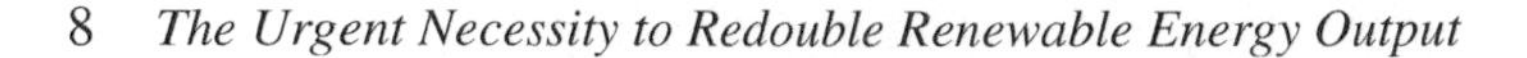

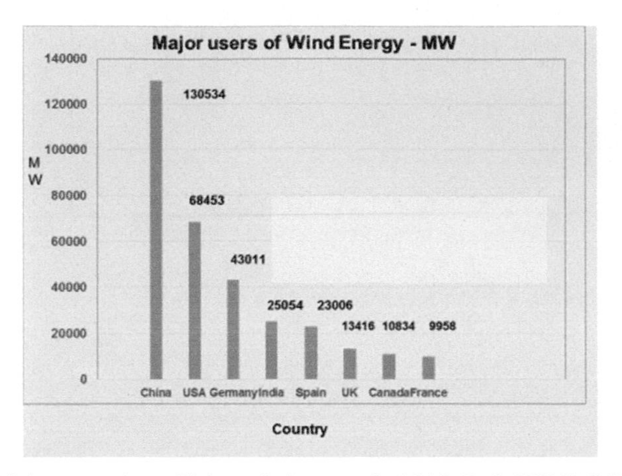

Figure 1.11 Major countries utilizing wind energy in 2015 (Ref.: WINDCATOR GLOBAL INDEX – Wind Power Monthly).

Table 1.1 W E - Cost breakdown all figures in $/kW

	Onshore	Offshore
Turbines	1400	2100
Foundations	160	600
Roads/transport	65	100
Electrical	210	300
Grid connection	150	400
Finance/legal	65	400–600
Total	2200	4000

Onshore O&M costs $ 8–16/MWh; Offshore $ 20–40/MWh. Ref.: David Milborrow – UK Consultant.

Considering the cost of wind turbines, Table 1.1 below shows the breakdown of the total cost of electricity produced by wind turbines on shore and offshore per kW, in 2015.

1.2.2 PV Technology

Progress in this area of renewable energy has exceeded all expectations. Nowadays, there is no country in the world now without some PV applications no matter how small the country or how small the application. It is truly a revolution in manufacturing, innovation, cost reduction and efficiency improvement (Figure 1.12).

Figure 1.12 (**a**) The 80 MW Okhotnykoyo Solar Park in the Ukraine, as a whole Ukraine has 13,551 MW with one array of 400 MW, and plans to have 50% of their electricity generated from PV by 2020; (**b**) Japan has added 537 MW of capacity since February 2014; (**c**) The Philippines has a renewable energy generation capacity of 5 GW.

In terms of technology several advances have been made during the last 3 years, for example:

- Magnolia Solar demonstrates has produced high performance coated glass which consists of coating the glass with a new class of materials consisting of porous silicon dioxide nanorods. The reflection losses at the glass–air interface have been reduced from approximately 4% to less than 1%. At large angles of incidence, the reflection losses have been reduced from over 25% to less than 5%.

- Flexible thin-film photovoltaic modules, have been manufactured by Ascent Solar, the EnerPlex Kickr IV – can be used on the beach, in picnic, and on the car dashboard – it is both light and strong. Plastic cells are much cheaper and flexible. A UK new study has shown that even when using very simple and inexpensive manufacturing methods, where flexible layers of material are deposited over large areas like cling-film – efficient solar cell structures can be made.

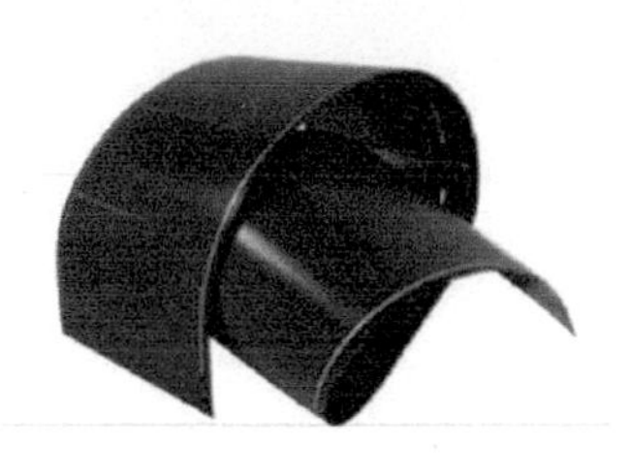
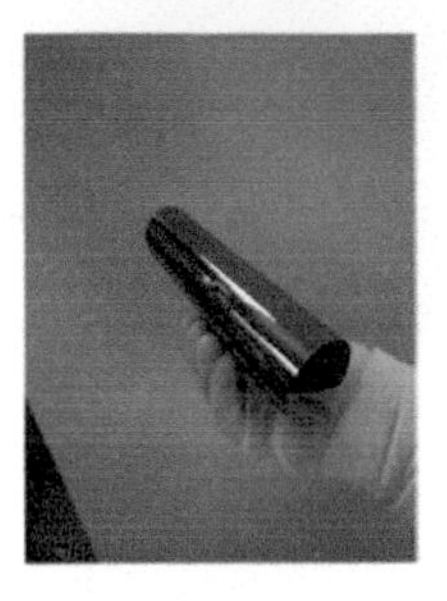

- Trina Solar has the highest efficiency record for the p-type mono-crystalline silicon solar cell on an industrial Cz wafer, which integrates advanced technologies including back surface passivation and local back surface field, reaching an efficiency of 21.40% (156 × 156 mm^2). In 2014, Trina produced 3.5 GW of PV, according to UK-based market research firm GlobalData, which represents 7.9% of global production. (Ref.: Global Energy World 19/11/2014)

- The Belgian solar glass manufacturer Ducatt NV and the US coating solutions provider 3 M, unveiled in 2015 a new water-based antireflective coating for solar modules. This increased the module's output by 3–5%. The coating was manufactured with Ducatt's 2 mm toughened flat glass and an environment friendly production process. (Ref.: PHOTON Newsletter 26 October 2015)

- Germany's Fraunhofer Institute for Solar Energy Systems ISE announced in September 2015 it has achieved a 25.1% efficiency for a both-sides contacted silicon solar cell.

- Trina Solar in December 2015 reported it has developed p-type monocrystalline silicon solar cell with an efficiency of 22.13%. $156 \times 156\,\text{mm}^2$. This compares with 25% the cell of $2 \times 2\,\text{cm}^2$ developed by Martin Green a long time ago.

- First Solar records 17.0% efficiency for a CdTe solar module, Cell efficiency of 20.4% in 2015.

- Boeing has set a PV Efficiency World Record, of 37.8% ground-based solar cell without solar concentration using multi-junction of two or three materials, reported by Global Energy World in 2013.

- The Energy Department at National Renewable Energy Lab announced in June 2013, a world record of 31.1% conversion efficiency for a two-junction solar cell under one sun of illumination. (AM 1.5, 1000 W/m^2 made of a gallium indium phosphide cell atop a gallium arsenide cell of 0.25 cm^2).
- SolenSphere Renewables produced a cell of 40% efficiency and when combined with the thermal energy captured by a parabolic concentrator, it approaches a total electrical and heat energy efficiency of 72%. This is cost effective and has cell area reduction of 1000. (Ref.: World of Renewable 12/9/2012)

The University of Bahrain and BAPCO has installed 5 MW of PV in Bahrain (Figures 1.13 and 1.14).

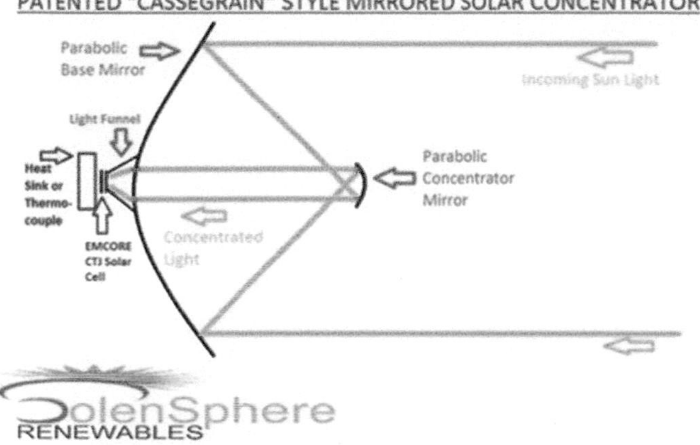

Figure 1.13 Various advances in PV technology.

Figure 1.14 Recent progress in PV utilization.

China installed more than 14 GW in 2014.

UK reaches 8.31 GW of PV capacity in 2015.
(Ref.: Photon Newsletter 27 November 2015)

USA has installed 13.48 GW in 2015.

Germany has installed a total of 2015 39.553 GW to date.

In China the total installation by September 2015 was 37.95 GW.

19 rooftop PV systems with a capacity of over 30 MW will be installed at the port of Jebel Ali, Dubai, UAE. The PV systems will cover one third of the port's electricity needs and will be installed on parking sheds and in the surrounding parking areas. It will be ready in 2016. (Ref.: PHOTON Newsletter 18 September 2015)

The total installation of renewable energy in India will have exceeded 4.1 GW in 2015.

Prime Minister Narendra Modi has set a target of 100 GW by 2022.

More than 1,463,867 rooftop PV systems, representing 4.47 GW of installed PV capacity, have been installed in Australia up to September 2015.

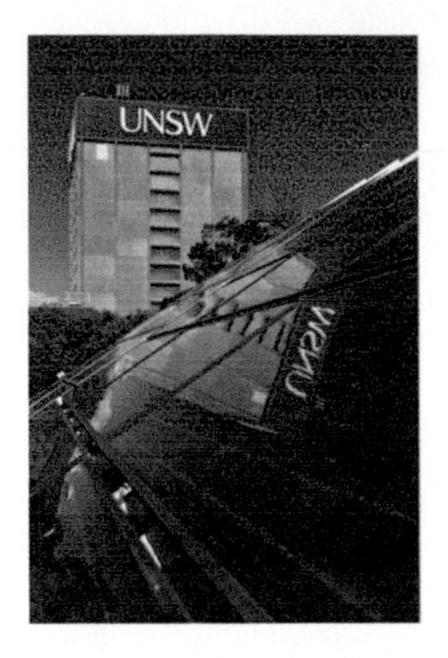

The Netherlands has reached 1.32 GW of PV capacity in 2015.

Switzerland has reached 1.35 GW December 2015.
(Ref.: PHOTON Newsletter 13 January 2016)

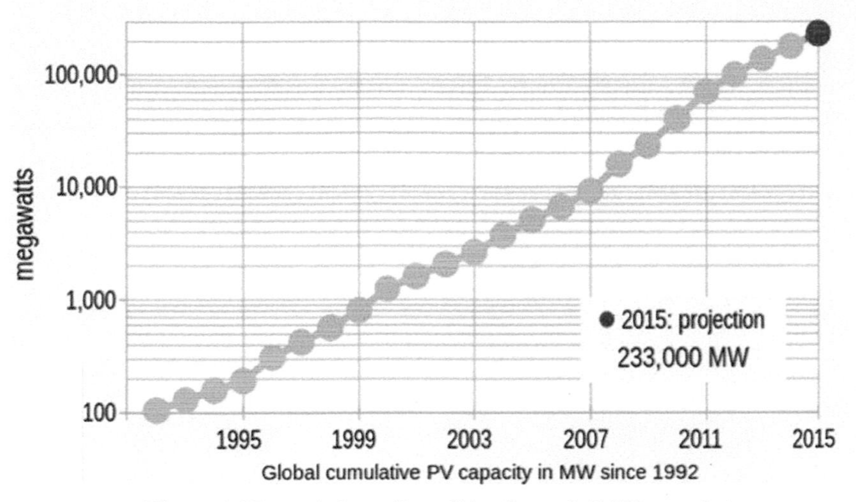

Figure 1.15 Ref.: https://en.wikipedia.org/wiki/Photovoltaics

Over the last 25 years the yearly production growth of PV has been more than 30% per annum (Figure 1.15).

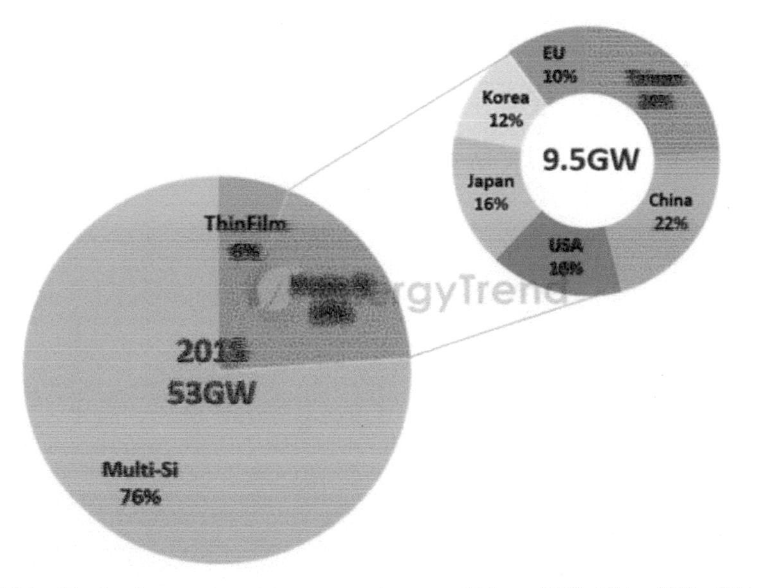

Figure 1.16 Production per country as percentage are: Taiwan 24%, China 22%, USA 16%, Japan 16%, Korea 2%, and EU 10% (Ref.: Energy Trend October 2015).

Figure 1.16 shows that the bulk of PV production in 2015 was of Multi Crystalline Silicon cells, of which Multi Si = 53 GW, Mono Si = 9.5 GW; while Thin Films were approximately 6% with 4.2 GW.

In terms of cost, Figure 1.17 shows the progressive decrease in PV cost from 1977 to date. The pay back of any PV system (1 kW–1 MW) is between (US$ 2.5–0.50).

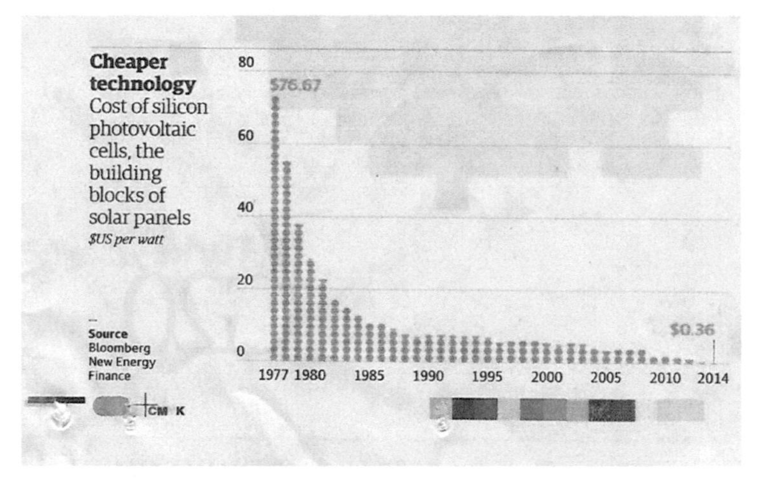

Figure 1.17

Fraunhofer Institute in their report: Photovoltaic Report 17 November 2015.

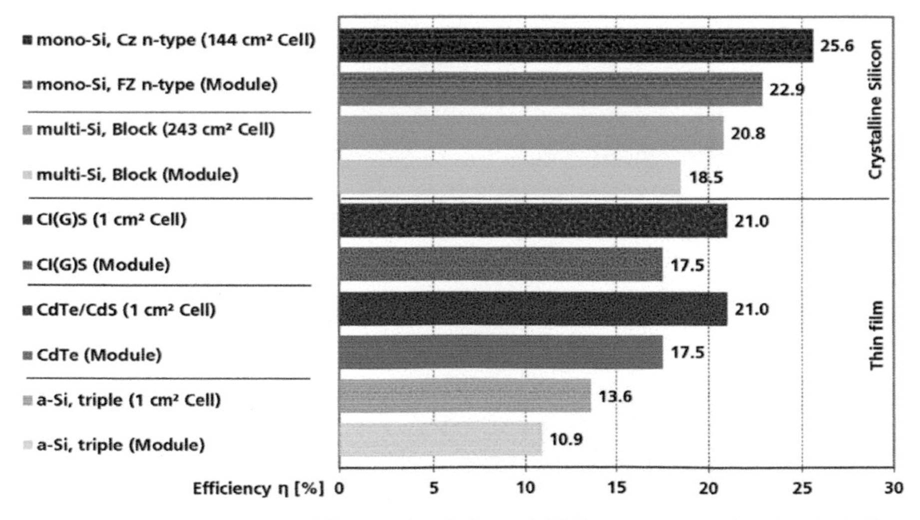

Figure 1.18 Efficiency comparison of technologies.

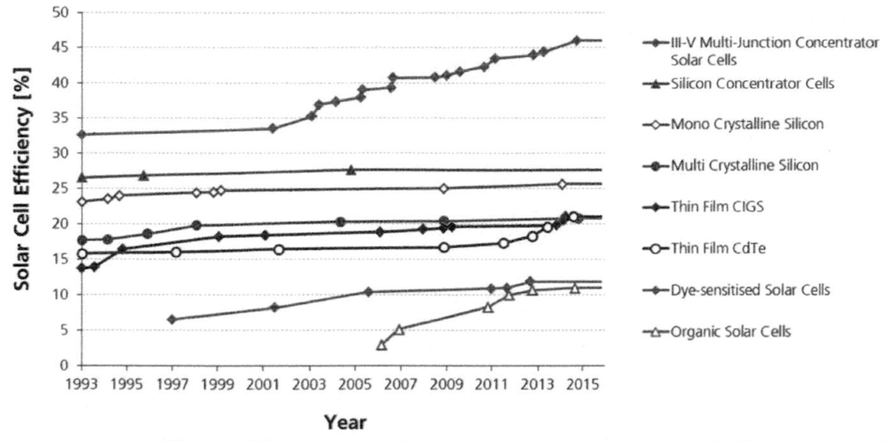

Data: Solar Cell Efficiency Tables (Versions 1–46), Progress in Photovoltaics: Research and Applications, 1993–2015.
Graph: Simon Philipps, Fraunhofer ISE 2015.

Figure 1.19 Development of laboratory solar cell efficiencies.

Stated of PV system (10–100 kW) in 2015 is US$1.5 per Watt. Generally the cost can be divided into: Cost of BOS including invertor is 52% and module cost is 48%.

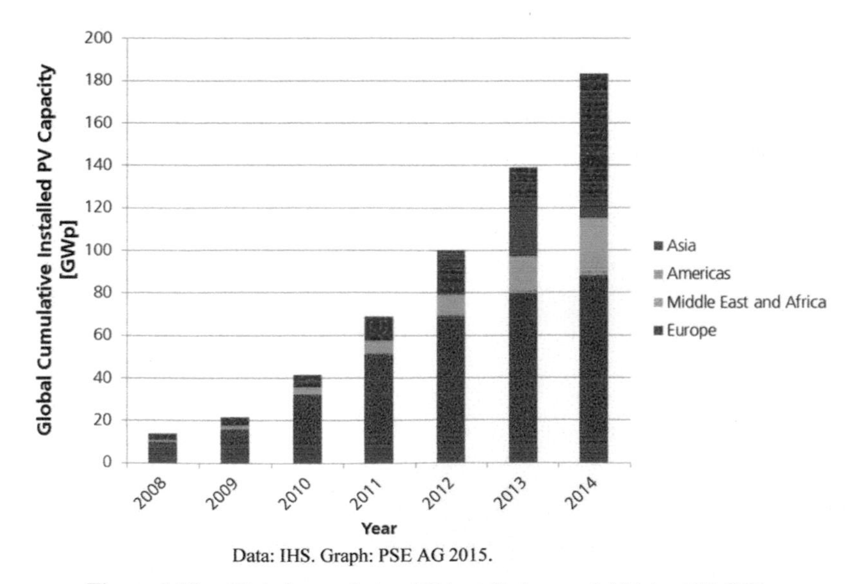

Data: IHS. Graph: PSE AG 2015.

Figure 1.20 Global cumulative PV installation until 2014 = 183 GW_p.

1.2.3 Hydropower, Ocean Wave, and Tide Energy

In a report written by David Appleyard, chief executive of *Hydropower & Dams*, '**Hydropower – leading global renewable energy capacity growth**', he states that: "New and authoritative figures from the REN21 Renewables Global Status Report 2015 indicate some 37 GW of new hydropower capacity was commissioned in 2014, increasing total global capacity by 3.6%. Indeed, hydropower forms one of a troika of technologies – the other two being wind and solar power – which have dominated new additions of renewable energy generating capacity over the last year. Globally, total installed hydropower broke 1,055 GW while worldwide hydro generation – which naturally varies each year with hydrological conditions – was estimated at 3,900 TWh in 2014, an increase of more than 3% on 2013 figures. And the top countries for hydropower capacity and generation remained China, Brazil, the United States, Canada, Russia, and India, which together accounted for about 60% of global installed capacity at the end of 2014. Of these, China alone commissioned almost 22 GW over the year, bringing its total installed capacity to 280 GW".

Roller-compacted concrete (RCC) dams are increasingly available, Jia Jinsheng, Vice president of Chinese National Committee on large dams stated in his article "7th International Symposium reviews state of the art of RCC technology, *Hydropower & Dams*, 22(6), 2015, 105": China now has more than 640 dams of RCC type with some reaching a height of more than 200 m. This has meant that the volume has increased so that now more than 22310^6 m^3 globally placed in RCC dams (Figure 1.21).

Figure 1.21 Longtan Dam – China RCC with height 217 m.

Dr Ian Masters chairman of Ocean Energy at the World Renewable Energy Congress outlined many applications in this area, one of the latest being the Osmosis Power Plant (OPP) which works on the use of a membrane through which water passes but salt cannot, resulting with fresh water on one side of the membrane and salt water on the other. Fresh water then travels to the salt water side due to concentration gradient, effectively pumping water which can then be used to power turbines. OPP are still largely experimental and as yet expensive. The major challenge is keeping the membrane clean and the requirement to locate the station near fresh and salt water (Figure 1.22).

Another system was explained by Dr Masters in the WREN International Seminar in Britain last November: **Ocean thermal energy conversion**.

This system uses the temperature gradient in deep, tropical waters to produce energy, which precludes its application in northern oceans. The system relies on the relatively small temperature gradient and the large volumes of water available (Figure 1.23).

Tidal Energy has been investigated based on barrages and lagoons where a large tidal range is available, e.g., Bristol Channel, UK. A reservoir is constructed within the tidal area which fills at high tide, at the change of

Figure 1.22 A 4-kW experimental OPP.

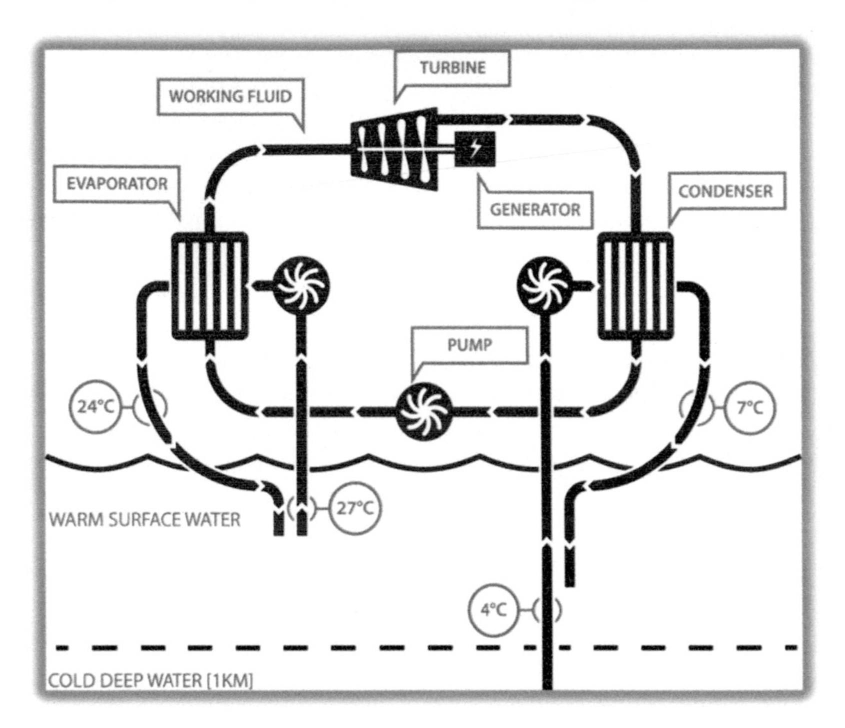

Figure 1.23 Ocean thermal energy power system. See http://www.bluerise.nl/technology

the tide it is emptied through a low head turbine, producing power. The River Thames Barrage, London, which has a tide height of up to 5 m and uses a low head turbine generates up to 50 MW.

Under water turbine and wave energy: Tides and water currents close to or near headlands or channels using under water turbines to generate electricity. Several devices are available (Figure 1.24).

Ocean energy is a natural phenomenon due to the tidal current which happens twice a day, every 12 h and 25 min. Tides can be thought of as waves; thousands of miles in length with the crest as high tide and the trough as low tide. Ocean tide is driven by the moon.

Power calculation from Tide Energy: If E = energy stored (J), m = mass of water (kg)

g = acceleration due to gravity (9.81 m/s^2)

H = head (tidal range) (m), v = volume of water (m^3)

ρ = density of sea water (1025 kg/m^3)

A = area of water impounded (m^2)

Figure 1.24 Under water turbine.

$Q = v/t$ = volume flow rate (m^3/s)
t = time (s), P = power (watts)

Potential energy stored:

$$E = mgH = \rho vgH = \rho AgH^2$$

Potential power extracted:

$$P = E/t = \rho vgH/t = \rho gHQ.$$

According to Betz Limit, only 59% of E can be extracted.

Therefore, the instantaneous power for tidal currents (stream): $P_s = 1/2\rho Av^3$.

Total theoretical UK resources are:

Tidal stream, 32 GW; Tidal range (barrage), 45 GW; and Tidal range (lagoon), 14 GW.

UK electricity power generation daily is 60 GW

Figure 1.25 Diagram for calculating the water movement in ocean power system.

1.2.4 Solar Thermal Applications

Desalination is the major application of Solar Thermal. Many countries do not have potable water for drinking which means that desalination is of prime importance. Table 1.2 shows some of the major desalination systems in operation.

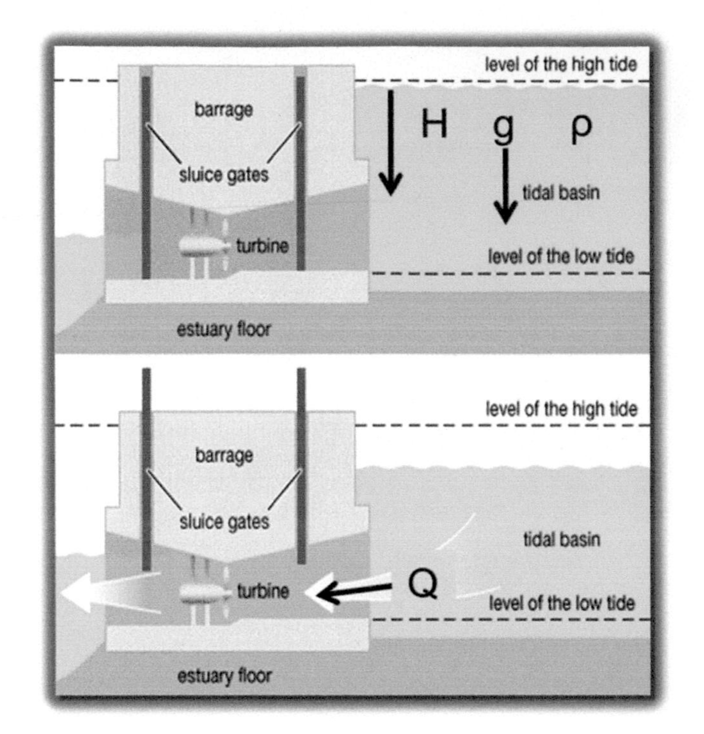

Figure 1.25

Table 1.2 Yield and energy consumption per process, reverse osmosis (RO), multi-stage system (MS), Multi-effect desalination (MED)

Process	Seawater (m^3) for 1 m^3 Water	Reject (m^3) for 1 m^3 Water	Thermal Energy MJ/m^3 water	Electrical kWh/m^3 Water
RO	22.9–1.7	11.9–0.7	—	3.5–5.5
MS	10.0–5.0	9.0–4.0	250–330	3.0–5.0
MED	5.0–2.9	4.0–1.9	145–390	1.5–2.5

1.2.4.1 Solar water heaters

Gone are the days when the only application of Solar Thermal was the ubiquitous solar heater seen on rooftops around the world. This is a comparatively low technology application. Nowadays the most efficient system is the evacuated tube collector where losses are low and cleaning requirements are minimal. There are expected to be more than 100 million solar water heaters in the US by 2050 (Figure 1.26).

The main source of energy is direct solar radiation (Figure 1.27).

Figure 1.26 Various water heater collectors in the market.

Figure 1.27 shows the solar radiation intensity in various locations in the world.

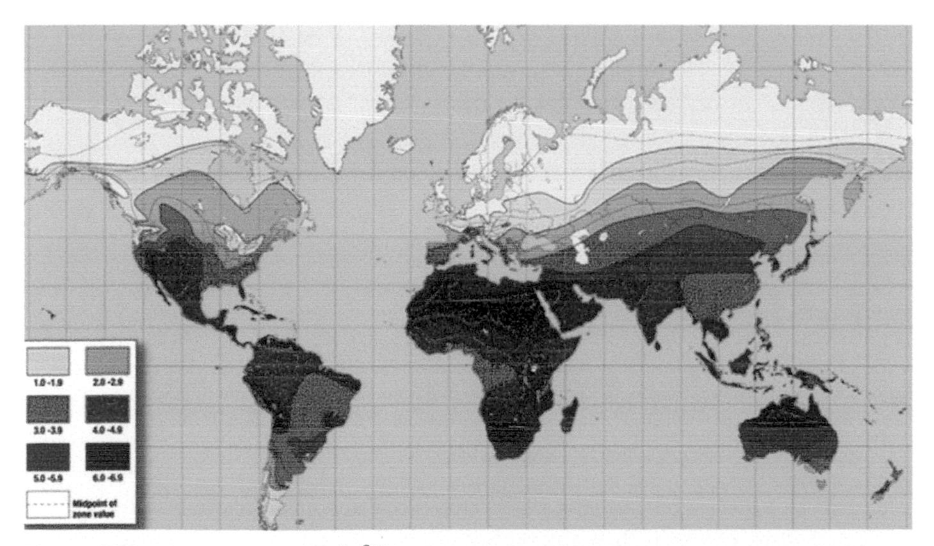

Figure 1.27 Radiation in kWh/m^2/day. Source: http://howto.altenergystore.com/Reference-Materials/Solar-Insolation-Map-World/a43.

Figure 1.28 Moroccan CSP in the south.

Concentrating Solar Power (CSP) is becoming very popular due to its combined solar storage facilities. There are many systems in operation in around the world, for example in southern Morocco a system of 160 MW – Noor1. Moroccan hopes to achieve 42% of its power from renewable energy by 2020 (Figure 1.28).

There are four types of CSP in operation at present; Fresnel Lens; Central Receiver; Parabolic Concentrator; and Stirling Engine. In 2015, there was 11,500 MW CSP in operation, 2100 MW in Spain, and 1820 MW in the US.

1.2.5 Fuel Cells

A fuel cell is a solid-state electrochemical power conversion device that directly converts the chemical energy of a fuel into electrical energy in a constant temperature process. A fuel cell comprises an electrolyte and two electrodes, the anode and the cathode, both of which are electronic conductors. The electrolyte which separates the two electrodes acts as an ionic conductor and does not allow electron flow through it (Figure 1.29).

In 2014, Ballard Company had powering 27 fuel cell buses manufactured by their partner Van Hool NV and operating in the Cities: Oslo, Norway

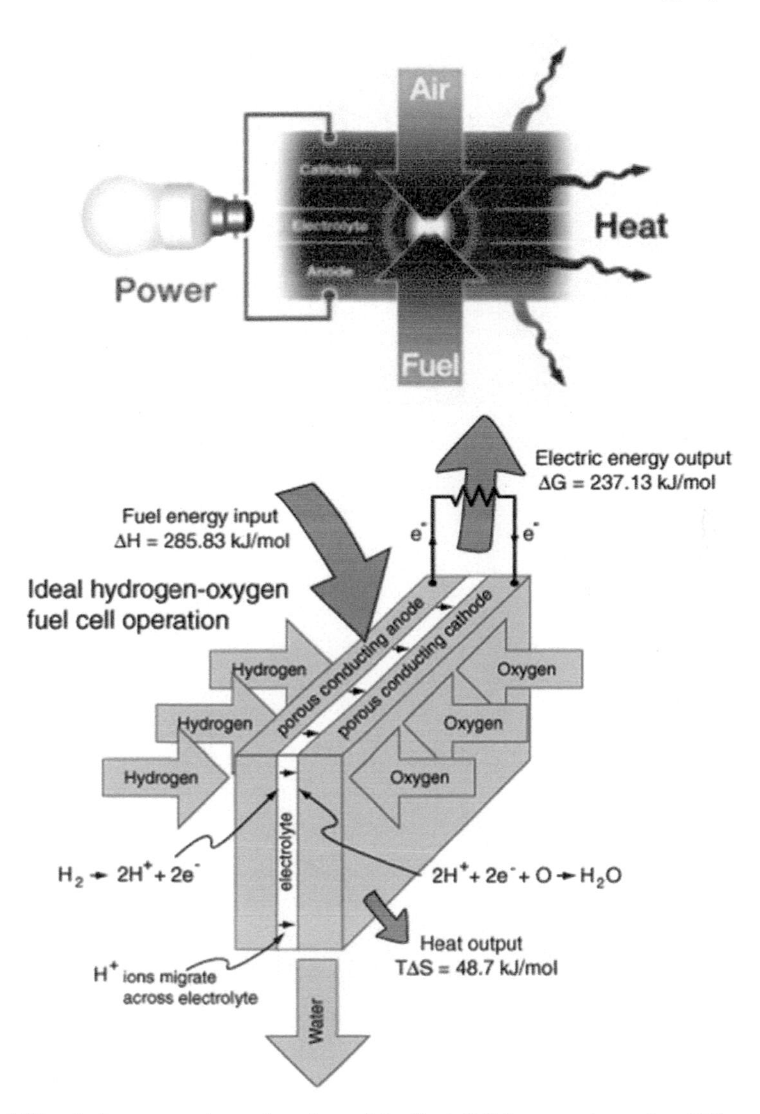

Figure 1.29 A diagrammatic explanation of the Fuel Cell in operation using different fossil fuels, deriving hydrogen and water from petrochemical fuel and gases (methane, ethanol, methanol, carbon monoxide, hydrides).

(5 buses); Cologne, Germany (2), San Remo, Italy (5), Flanders, Belgium (5) Aberdeen, Scotland (10), while a fleet of 20 buses that has been operating in Whistler, Canada since 2010 (Table 1.3).

Table 1.3 The various types of fuel cells

	Low Temperature			High Temperature		
Fuel Cell Type	Alkaline	Proton Exchange Membrane	Direct Methanol	Phosphoric Acid	Molten Carbonate	Solid Oxide
Symbol	A	PEM	DM	PA	MC	SO
Operating Temperature°C	50–200	50–100	60–120	180–220	650	500–1000
Efficiency	35–60	40–60	40–60	40	50	45–65
Applications	Military and space	Transport, portable power, black up power and small generation	portable power	transport, distributed generation	Electric Utility and Large distributed generation	Electric utility, auxiliary power and large generation

1.2.6 Biomass, Biogas, and Waste to Energy

One of the easiest ways of obtaining biogas is the process which is called Anaerobic Digestion. This is a simple process of creating optimum conditions for bacteria growth. As time goes by the feedstock is digested by these bacteria that in turn generates two byproducts – digestate and biogas, (Ref.: REA Web) (Figure 1.30).

Some of the large scale biogas digesters are shown below in Figure 1.31

There are two major methods to produce Biofuel: the first uses root crops such as potatoes and sugar beet, or cereals such as wheat and maize, sugar beet. The product is bioethonal.

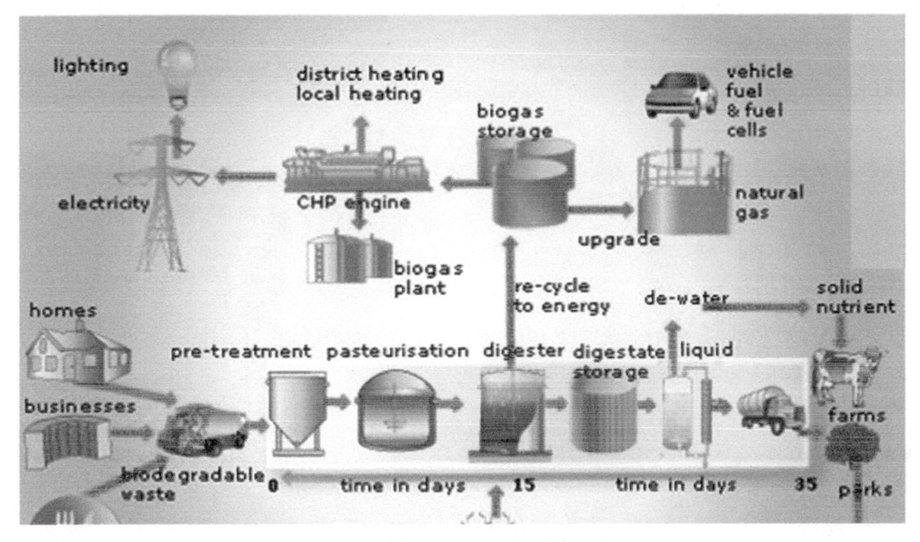

Figure 1.30 Anaerobic digestion.

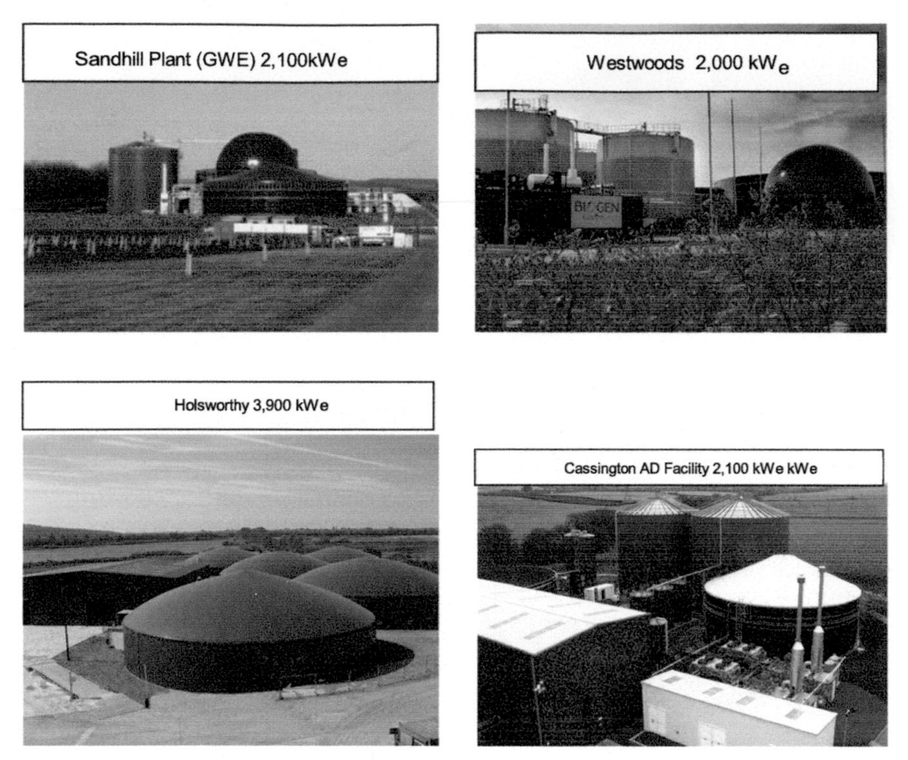

Figure 1.31 Four large scale biogas digesters.

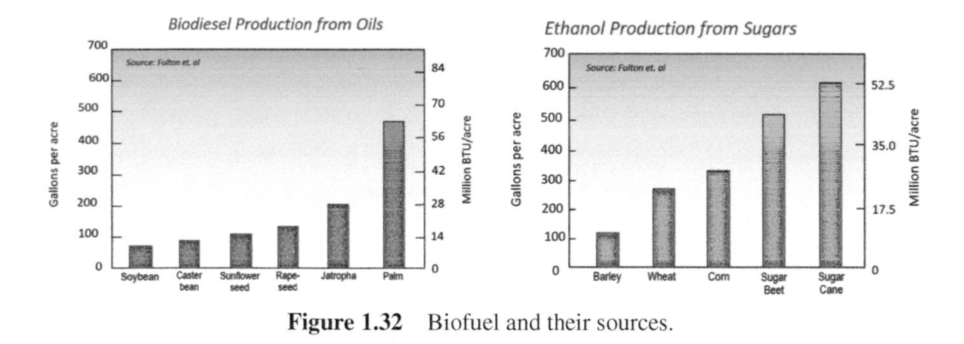

Figure 1.32 Biofuel and their sources.

The other major method uses oil seeds as the feedstock: OSR, sunflower, soybean, palm or jatropha, the product in this case is biodiesel (Figure 1.32).

However, with regard to biofuel production there are ethical issues. WREN was part of a team organized by Nuffield Council which agreed upon the following:

1. biofuels should not be developed at the expense of human rights,
2. they should be environmentally sustainable and
3. they should contribute to a reduction of greenhouse gas emissions
4. their sale should adhere to fair trade principles and
5. the costs and benefits of biofuels should be distributed in an equitable way.

Ref.: April 2011, Nuffield Council on Ethics.

Many believe that one of the most rewarding sources of biofuel is algae fuel. The algae can be grown in both fresh and salt water and yields a large amount of biofuel in a reasonably short period of time. Although it is costly at present future technological advances will reduce the capital and operating costs and produces more fuel per unit area than other biomass sources (Figure 1.33).

Figure 1.33 Australia's first integrated demonstration plant converts algae into green crude.

Table 1.4 Oil production from various crops

Gallons of Oil per Acre per Year	
Corn	18
Soybeans	48
Safflower	83
Sunflower	102
Rapeseed	127
Oil Palm	635
Micro Algae	500–1500

The renewable energy fuels company, Muradel, has launched Australia's first demonstration plant producing 30,000 l of crude oil a year at Whyalla in South Australia. This will be upgraded to a commercial plant, producing 500,000 barrels of green crude a year by 2019 – enough petrol and diesel to fuel 30,000 vehicles for a year. This will be upgraded to a commercial plant, producing 500,000 barrels of green crude a year by 2019 – enough petrol and diesel to fuel 30,000 vehicles for a year (Table 1.4). (Ref.: Direct communication, November 2014)

1.2.7 Waste to Energy (WTE)

A new important source of renewable energy is that of the incineration of municipal and agricultural wastes to produce electricity. Many countries continue to prefer dealing with these waste products by means of landfill despite the well-known disadvantages of gas leakage from the landfill sites which is harmful to humans and means that the land cannot be used for building after the site is full and it occupies vast tracts of land. Figure 1.34 shows methods of processing waste.

1.2.8 Geothermal Energy

Geothermal energy involves drilling into earth's core and tapping the hot steam stored at various depths. 67 countries now having geothermal applications, the US is in the lead (16600 MWh), Philippines (9646 MWh), Indonesia (9600 MWh), Mexico (7071 MWh), and New Zealand (7000 MWh), while several South America countries and Iceland also utilize geothermal for electricity generation (Table 1.5).

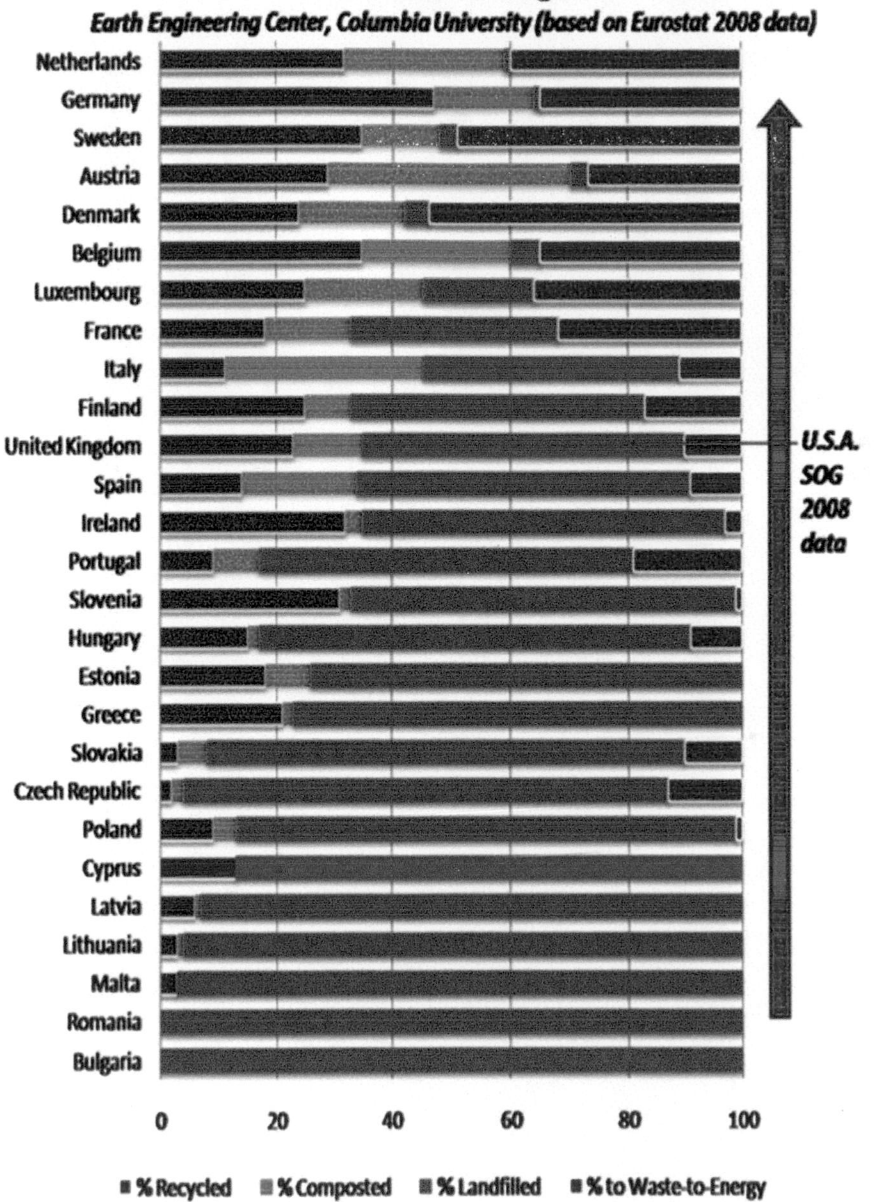

Figure 1.34 Percentage of WTE usage in some countries.

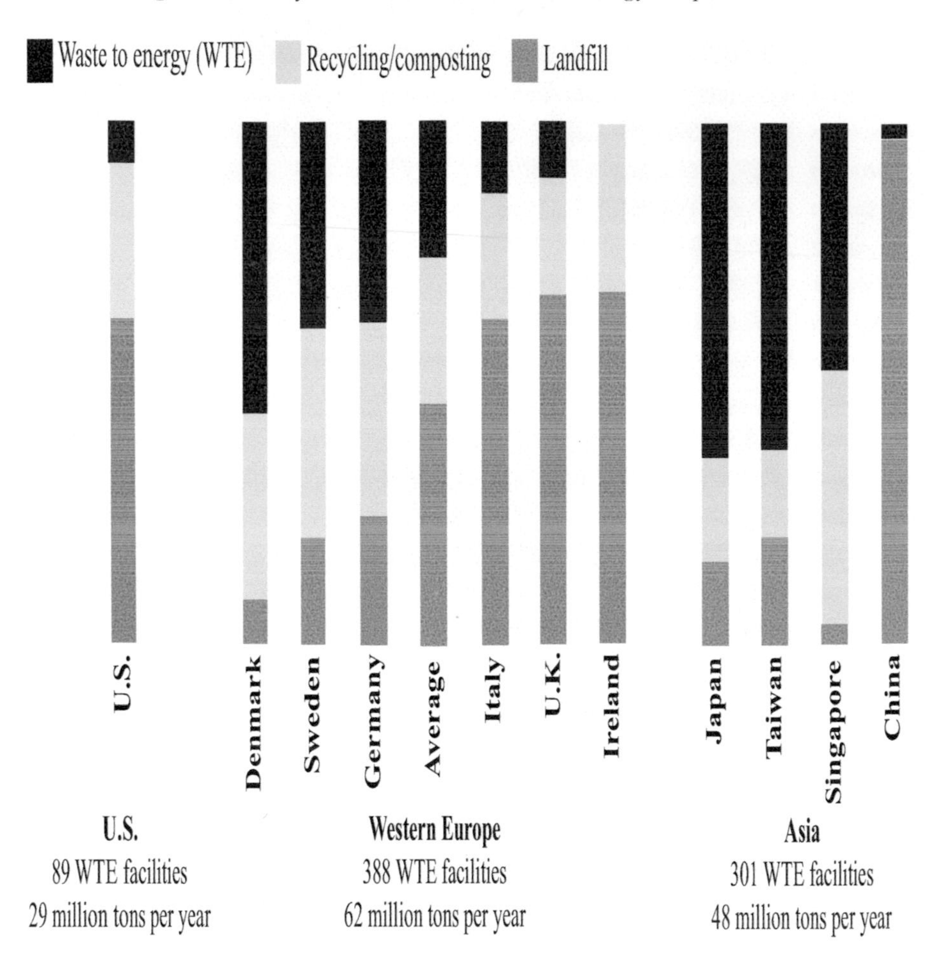

Table 1.5 Total worldwide installed capacity from 1995–2015

Year	Installed Capacity MWe	Produced Energy GWh
1995	6832	38035
2000	7972	49261
2005	8933	55709
2010	10897	67246
2015	12635	73549
2020	21443	

Ref.: Proceeding of World Geothermal Congress 2015.

	Year					
	2005		2010		2015	
Country	MW	GWh	MW	GWh	MW	GWh
Europe	1123	7209	1643	11371	14821	3385
Africa	136	1088	209	1440	2858	1601
America	3911	25717	4565	26803	26353	8305
Asia	3290	18903	3661	23127	22084	6712
Oceania	441	2792	818	4506	7433	1440
TOTAL	8903	55709	10897	67246	73549	21443

Below are two examples of geothermal plants.

Chile 12 MW Geothermal Well, Tol-4, was drilled to a depth of 2,300 meters produces high power steam, enough to supply US a geothermal plant.

The potential of US 45,000 homes with electricity Geothermal is 3,000,000 MW. At present the cost is less than 10 cent/kWh.

1.3 Conclusion

The progress in renewable energy technology is very encouraging. The same cannot be said of implementation, more commitment from policy makers and governments is needed to offset the danger from climate change. Nations must commit themselves to use renewable energy NOW, the time for protracted self-interested discussions has passed. Climate change is destroying crops, creating health hazards, and causing thousands of deaths and destruction through famine and floods. Fossil fuels can be used for other applications instead of transport and electricity generation.

In brief:

- Wind energy generation cost US$ 2000 per kW and the payback is 3 years. This is of course when wind speed 5 m/s or more.
- Photovoltaic technology can be used in a wide variety of applications from transport, space travel, electricity generation, heating and cooling, desalination and agriculture. Costs have dropped to US$ 0.50 per Watt. Efficiency has improved and new technologies are developed every day, for example thin films such as CIGS 21.7%, and passivated emitter rear contact solar cells – PERC – Solar World Company announced in January 2016 an efficiency of 22.04%. Trina in China also has announced their crystalline cells have reached an efficiency of 21.4%. The payback of any system 10–100 kW is less than 2 years.
- As for biomass and biogas and WTE, the recent achievements are remarkable especially in air traffic fuel. Many countries are investing heavily in burning municipal solid waste (MSW) than rather in landfill processes. Following intensive research it is now possible to generate electricity from cultivating algae and using it either as liquid fuel as in some cities in France or burning it as solid fuel.
- Many more countries are investing in solar thermal power plants – concentrating solar power (CSP).
- Geothermal application has progressed with a more than 14% increase in the number of systems in use than in the previous year.

The EU is to be congratulated in committing every member country to generate 20% of their electricity from renewable energy by 2020. Spain now generates more than 50% of their electricity from renewable energy; Germany has pledged that by 2035 65% of their electricity will be generated from renewable energy; remarkably Denmark has pledged that 100% of their electricity will be generated from renewable energy by 2050. While many non-EU countries have also pledged that substantial increases will be made in the percentage of their electricity coming from renewable energy in the next decade.

The way forward now is to increase the investment in energy conservation, to improve the efficiency of all renewable energy devices, and to provide incentives to importers or to the users such as feed-in- tariffs (FiT) and financial loans to make a level playing field with all other energy sources.

World Renewable Energy Network predictions for the next 15 years are:

1. Electric cars will have 50% share of the market.

Already many manufacturers of cars are heavily investing in electric car production, see below the two examples.
Hybrid Solar Car by Ford, capable to cover 42 km/l. It uses concentrated PV with tracking east-west system. Each car will reduce CO_2 by 4 Tons a year.

Tesla Motor Company Delivered. In 2015 50,580 Electric Cars with all-wheel drive and a 90 kWh battery providing 257 miles of range.

2. Fuel cells applications will increase to a level so that 50% of all electric power stations will operate by fuel cells while in the heavy transport sector 50% of the vehicles and trains will run on fuel cells.

An example here showing the world's largest fuel cell park in South Korea in The Gyeonggi Green Energy fuel cell park, located in Hwasung City, built by POSCO Energy. It was completed in the first quarter of 2014 consisting of 21 modules each 2.8 MW, another one will be constructed in Seoul City of 19.6 MW. The largest fuel cell park in the world is providing 59 MW of electricity and District Heating to 45,000 homes, covering only 5.1 acres. Presently worldwide exist more than 50 locations having total of 300 MW.

3. Photovoltaic applications will cover 50% of all newly built houses in the world. Improvement in efficiency, less material, and reliability make PV a viable option for electricity generation in buildings, see the two examples.

2

Future Concepts in Solar Thermal Electricity Generation

Ing Marc Röger

DLR German Aerospace Center, Institute of Solar Research,
Plataforma Solar de Almería, 04200 Tabernas, Spain

2.1 Introduction to Concentrating Solar Power (CSP)

Concentrating Solar Power (CSP) or Solar Thermal Electricity (STE) comprises all technologies that concentrate the solar light by mirrors or lenses to generate heat which can be converted to electricity by a conventional power block including turbine and generator. The collected heat can be optionally stored in a thermal storage and be recovered to produce electricity when needed, for example, during the peak demand in the evening, although there is no solar radiation. The CSP technology distinguishes from Photovoltaics (PV) which directly converts sunlight to electricity using semiconductors (Figure 2.1). On a photovoltaic cell, the photons create excited electrons and remaining holes which are separated to generate an electric voltage. The electricity is produced instantaneously and has to be consumed or stored in electrical storages. PV converts both diffuse and direct solar radiation, whereas CSP only concentrates the direct solar radiation. Consequently, CSP plants will be favorably located in the sunbelt countries.

Additionally to the generation of electricity, concentrating solar technologies can provide process heat to industrial processes or heat to produce chemical commodities or solar fuels. Concentrated solar light can also be used in concentrating photovoltaics (CPV). High-performance, higher-priced multi-junction solar cells with lab efficiencies up to 46% [1] may be used soon to convert the concentrated radiation to electricity.

During the recent years, the deployment of more than 150 GW of photovoltaic cells has led to significant cost reductions and grid parity in a lot of countries. However, as being a fluctuating renewable source, this positive

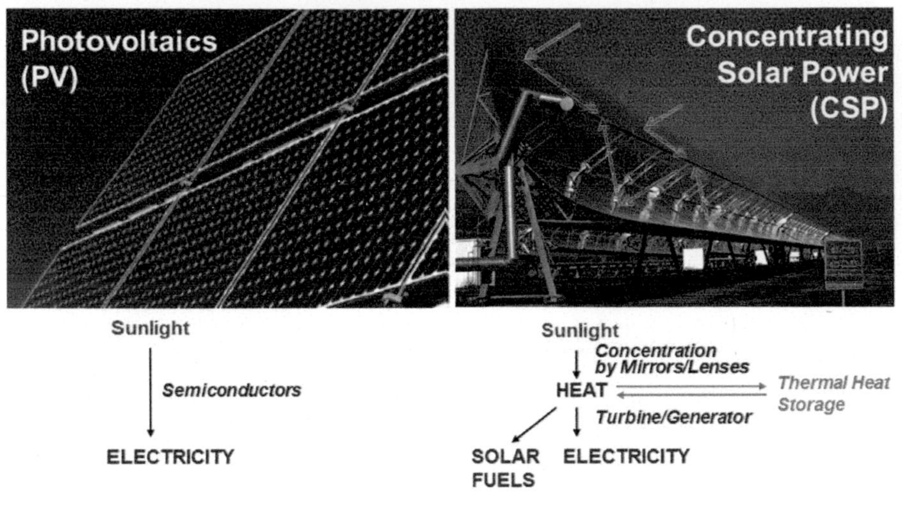

Figure 2.1 Conversion of sunlight: PV (left) and CSP (right).

development also creates issues regarding grid stability during moments with high demand and no solar radiation. Here, the CSP technology with its inherent capability to store the collected solar energy in low-cost thermal storage systems (compared to state-of-the-art electrical storages) can provide firm and flexible capacity. The deployment of CSP technologies with almost 5 GW in operation and 5 GW under development or construction worldwide lacks behind compared to PV. Most of the CSP plants are located in Spain, USA, and the MENA region. South Africa, Chile, China, and India are emerging. The CSP learning curve is still at the beginning and there is broad agreement that actual prices (about 14 € cents/kWh, 2015) will still fall significantly. Levelized electricity costs of CSP need not necessarily be competitive to PV systems, because CSP technologies have a much higher capacity value. Plants offering capacity are needed to stabilize electricity grids which have high shares of intermittent renewable energy sources such as solar PV or wind. Hence, CSP complements PV and wind. The International Energy Agency foresees that CSP will have a share of global electricity generation of 11% by 2050 (980 GW). The construction rate will increase after 2020. According to IEA, PV could produce another 16%, resulting in 27% solar energy of electricity produced worldwide by 2050 [2]. Worldwide renewable electricity production could be almost 80%.

This text presents future concepts for CSP plants which further reduce costs of renewable electricity generation.

2.2 State-of-the-Art CSP Plants

Different CSP technologies are shown in Figure 2.2. Parabolic trough collectors and linear Fresnel collectors are line-focusing technologies. Central receiver systems, also called solar towers and Dish systems are point-focusing systems. In general, point-focusing collectors can reach higher solar concentration factors and are used to drive processes requiring higher temperatures.

A parabolic trough collector tracks the sun over one axis and concentrates the radiation on the focal line. A vacuum receiver with selective coating absorbs the radiation and transmits it to the heat transfer fluid. Thermal oil is mostly used, having a temperature limitation of about 390°C. Advanced heat transfer fluids like molten salt, silicone fluids, and water/steam with higher operation temperatures are currently under investigation. The heat is transferred to a two-tank thermal storage using molten salts as storage liquid. The molten salt circuit transfers the heat to the water/steam circuit of a Rankine cycle. In a state-of-the-art parabolic trough collectors, the receiver moves together with the mirrors over the day, whereas in a Fresnel collector, the receiver is fixed. Stripes of mirrors on the ground track the sun on the receiver. This not only leads to lower costs of the Fresnel collectors compared

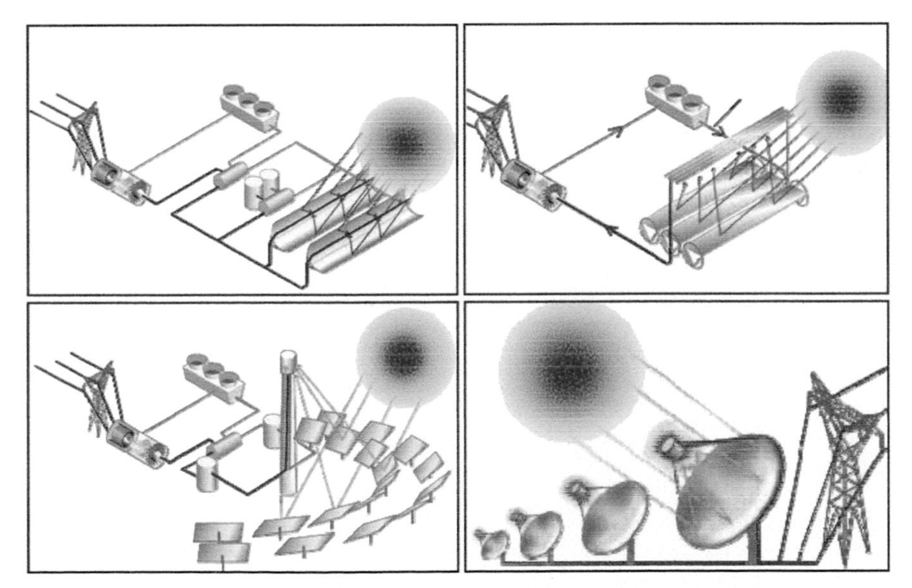

Figure 2.2 Concentrating Technologies: Parabolic trough collector, Fresnel collector, dish system, central receiver system (from top left to bottom right) [4].

to a parabolic trough system, but also to lower annual efficiencies. Typical solar concentration ratios for line-focusing collectors are 80 with working fluid temperatures between 200 and 400°C.

In central receiver systems, hundreds or thousands of heliostats track the sun over two axes. The sunlight is concentrated on a receiver which is located on top of a tower. High solar concentration factors of over 1000 suns can be achieved with typical operation temperatures of 565–1100°C. Dish systems are much smaller with power levels between 3 and 250 kW. They usually have high optical accuracy, yielding easily solar concentration factors of several thousands. The paraboloidal concentrator is tracked together with the receiver which often feeds a Stirling motor, maybe in future also micro gas turbines. Table 2.1 shows the characteristics of two state-of-the-art CSP plants being in operation. More information about the CSP technology can be found at the SolarPACES or ESTELA homepages [3, 4], for example.

2.3 Cost Structure of CSP Plants

In contrast to fossil fuel plants, a great part of the cost of a solar power plant incurs during the planning, construction and commissioning phase. After that, CSP plants almost produce electricity at very little cost. Hence, the annualized capital cost (CAPEX) is the cost driver of a CSP plant. The annualized CAPEX for an exemplary 100-MW central receiver system with 15-h storage is estimated to 84% [6].

Table 2.1 Example of two operational CSP plants [5]

	Parabolic Trough Plant	Central Receiver System
Name	Andasol-3	Crescent Dunes
Location	Aldeire, Spain	Tonopah, NV, USA
Developer	Ferrostaal AG	Solar Reserve, LCC
Land	2,000,000 m^2 (280 soccer fields)	6,475,000 m^2 (906 soccer fields)
Collector	510,120 m^2 aperture (71 soccer fields, 156 loops a 4 collectors, 150 m long, one-axis tracking)	1,197,148 m^2 aperture (168 soccer fields, 10,347 heliostats, each 115.7 m^2, two-axis tracking)
Receiver	HTF VP-1 in selective receivers 293°C → 393°C. Receiver length: 93.6 km	Molten salt in tubular receiver 288°C → 565°C. Tower height: 195 m
Turbine	50 MW_{el}	110 MW_{el}
Thermal Storage	two-tank, molten salt, 7.5 h capacity	two-tank, molten salt, 10 h capacity

Having a closer look at the CAPEX cost share, about half of the capital cost is caused by heliostat field and receiver. In order to reduce costs for STE generation, future technological concepts have to tackle these main cost drivers.

2.4 Common Features of Future Concepts

There is a variety of measures to reduce costs of future CSP plants. This text tries to identify the common features future concepts may have:

(a) *Higher concentrations* for *higher temperatures* and *efficient cycles* reduce solar field and receiver size and hence costs.
(b) *Dispatchability* increases the value of CSP electricity by offering dispatchable electricity.
(c) *Systems with reduced complexity* lead to capital and O&M cost reduction.

Further non-technological issues to reduce costs are economies of scale, like plant size scale-up, repetition of plants and component mass production, qualification, performance testing, and standardization to reduce the technological project risks and facilitate the bankability, and last but not least, a stable, reliable, and long-term regulatory framework. These non-technological issues are of highest importance for a successful deployment of renewable energy technologies, although not being the focus of this text.

2.4.1 Higher Concentrations, Temperatures and System Efficiencies

Solar concentrating technologies can provide different levels of solar concentration ratios. Future concepts might preferably have high-concentration ratios, being over 100 suns for line-focusing systems and over 1000 suns for point-focusing systems. These technologies easily can provide high-temperature heat with good solar thermal (collector) efficiency to power highly efficient thermodynamic cycles (Figure 2.3 (a)).

Figure 2.3 (b) illustrates the conversion of the collected heat to electricity. The maximum theoretical thermodynamic limit is known as Carnot efficiency, which increases with the maximum temperature of the thermodynamic process. Due to irreversibilities, the practically achieved efficiency is lower. Rankine cycles achieve typically between 35 and 45%, supercritical steam or closed-loop supercritical CO_2 turbines efficiencies over 50% at still moderate temperatures between 600 and 800°C. Combined cycle plants require much

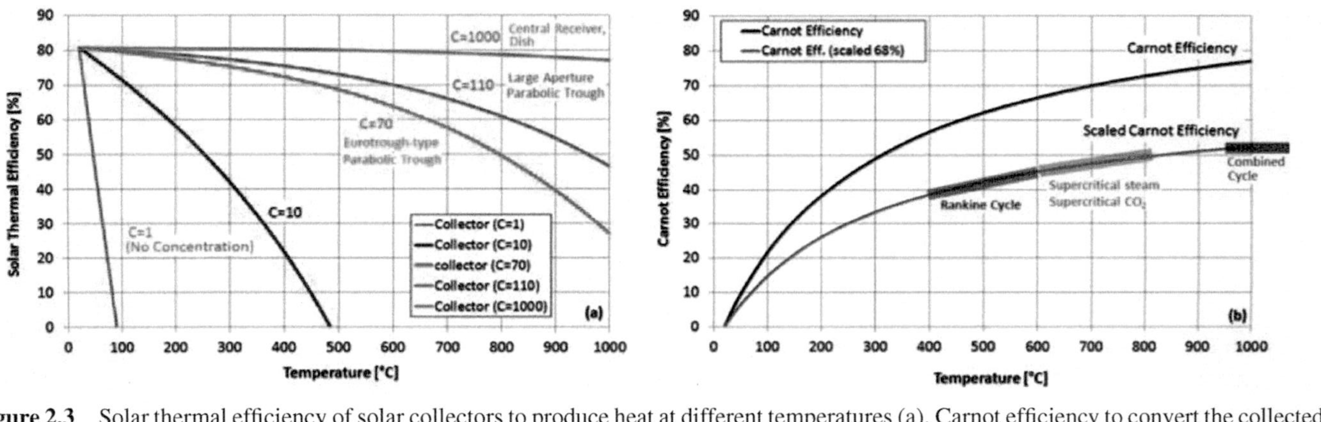

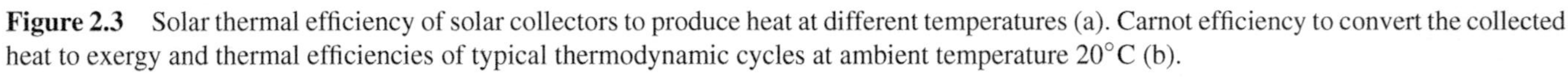
Figure 2.3 Solar thermal efficiency of solar collectors to produce heat at different temperatures (a). Carnot efficiency to convert the collected heat to exergy and thermal efficiencies of typical thermodynamic cycles at ambient temperature 20°C (b).

higher temperatures typically between 1000 and 1400°C, but compensate with efficiencies between 50 and 60%. High-temperature heat can be transformed to electricity with highly efficient cycles such as superheated steam, supercritical steam, supercritical CO_2 (closed Brayton), and combined cycles.

Figure 2.4 shows the total system efficiency as the product of the solar thermal collector efficiency (Figure 2.3 (a)) and the scaled Carnot efficiency (Figure 2.3 (b)). Each concentrating technology combined with the appropriate cycle has its optimum at a different maximum process temperature. Central receiver systems and large-aperture parabolic troughs for example are appropriate concentrator systems to achieve high efficiencies. Solar systems generating very high temperatures over 1000/1100°C have drawbacks regarding material issues and higher radiation losses. These drawbacks may be not paid off by the benefits of more efficient thermodynamic cycles. Moderately high receiver temperatures between approximately 600 and 1000°C for point-focusing systems and 550°C for line-focusing systems may be sufficient for not facing excessive material problems while maintaining good total system efficiency.

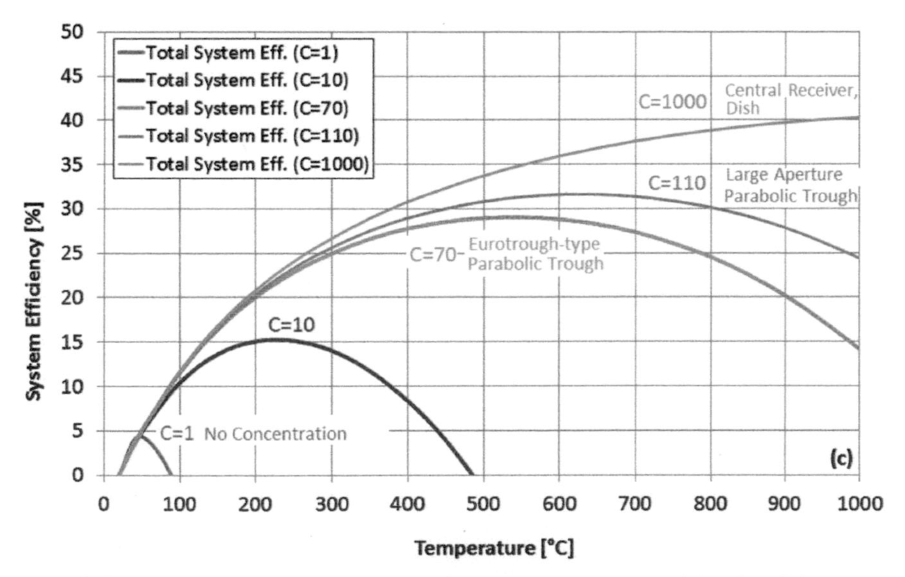

Figure 2.4 System efficiency (conversion of solar radiation to electricity) for different concentrating collector systems operating at different temperatures. Heat to electricity conversion using the scaled Carnot efficiency.

2.4.2 Dispatchability

The characteristic to easily and economically integrate a thermal storage system into a CSP plant converts the CSP technology from a fluctuating renewable energy source to a firm and flexible capacity of the energy supply which helps to increase grid stability. A thermal energy storage is much cheaper (current costs about 40€ /kW_{th}) than a state-of-the-art electric storage (pump-storage, battery, and supercapacitor) and has high efficiencies over 95%.

Hence, on the energy system level, CSP both enables a higher penetration for fluctuating renewables like wind and PV and reduces energy system costs by avoiding excess capacities of wind and PV which may lead to curtailment of energy. The capacity value of CSP electricity is due to its dispatchability higher than PV, see also Denholm et al. [7].

Depending on the electricity load curve and the type of service a CSP plant should provide, the design is adapted. Different combinations of solar field, storage, and turbine size permit different services. Intermediate load during the daytime can be covered by a CSP plant with a small thermal storage to buffer cloud passages. Base load is covered by a large thermal storage and a smaller turbine which runs 24 h a day. Peak load is produced by a design including a large storage which accumulates heat over the whole day and discharges during few peak hours to run a large turbine. The delayed intermediate load case is given, if the energy of the sunshine hours is shifted to the evening hours, for example. Here, a medium-sized storage with a medium-sized turbine is used.

Figure 2.5 shows an exemplary load curve and an electric dispatch curve for a summer week in a virtual energy system of the year 2040 without unit commitments. The CSP plants preferably deliver electricity to the system after sunset or when there is no wind energy.

Physical and chemical processes serve to store the heat: Sensible heat storages use either liquids (e.g., molten salts, metals, and water–steam) or solids (e.g., moving particles, rocks, and concrete). Latent heat storages use phase change materials, which have ideal characteristics for direct steam generation and Rankine processes. The heat transfer from the working medium to the storage can be either a direct-contact heat exchange or realized by heat exchanger. The energy storage in a chemical, reversible reaction (e.g., dehydration of $Ca(OH)_2$ or decomposition of H_2SO_4) has a higher energy density and the heat can be stored long-term without heat loss.

2.4.3 Systems with Reduced Complexity

In CSP technology, there is still potential to reduce system complexity. Innovations to reduce these complexities should be pushed into the market by

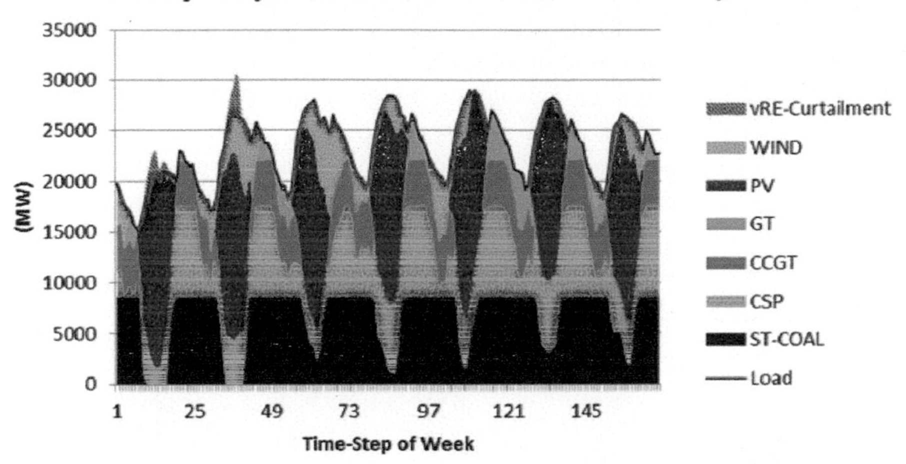

Figure 2.5 Exemplary load curve and electricity dispatch for a summer week in a virtual energy system (year 2040) without unit commitment, including wind, PV, gas turbines (GT), combined cycle gas turbines (CCGT), CSP, and coal power plants [8].

actual R&D activities. The state-of-the-art parabolic trough plant with molten salt storage, for example, uses three different fluids: the oil-based heat transfer fluid of the solar field, the molten salt circuit of the thermal storage, and the water–steam circuit of the Rankine cycle. A future concept could preferably use only one medium for receiver and storage system. Also, components like heliostats and receiver layouts may have potential for simplification. Other examples to reduce complexity and costs are systems which work under or near atmospheric pressure. This simplifies the requirements for construction and usually saves a lot of materials due to reduced wall thicknesses. Section 2.5 lists some examples for systems with reduced complexity.

2.5 Examples of Future Concepts

The following concepts meet most of the features presented in Section 2.4. They use high solar concentrations, high temperatures, and hence profit from highly efficient cycles. They all have a thermal storage for increased dispatchability. Their complexity is reduced due to the fact that there is mostly only one medium for receiver and storage, and the presented solar receivers and storage systems are non-pressurized or work near atmospheric pressure.

2.5.1 Solar Tower with Liquid Heat Transfer Fluid and Storage

A state-of-the-art heat transfer fluid in central receiver systems is molten salt [3, 9] (Figure 2.6). The operating temperature of solar salt ranges approximately between 220 and 565°C. Outside this range, it either freezes, causing problems to melt it again, or it decomposes. Hence, a proper drainage of the tubing system after shutdown of the receiver or a proper heat tracing is necessary. Equally important, highly efficient thermodynamic cycles using temperatures above 565°C cannot be driven with solar salt. For these reasons, metals are interesting candidates which are under investigation [10].

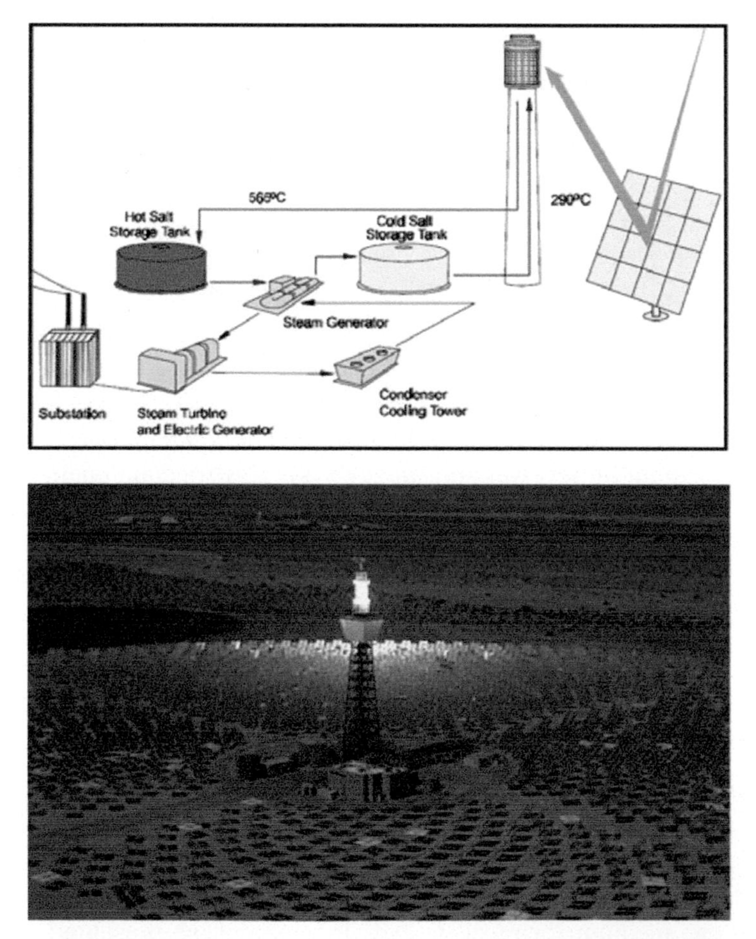

Figure 2.6 Central receiver system with molten salt as receiver and storage heat transfer fluid. Solar Two configuration [3, 9].

Table 2.2 Liquid heat transfer fluids and their temperature range

Heat Transfer Fluid	Minimum Temperature (°C)	Maximum Temperature (°C)	Remarks
Solar Salt	220	565	Actual state-of-the-art fluid, low max. temperature
Sodium	98	890	Highly reactive, exothermic reaction, very good heat transfer coefficients
Lead-Bismuth	124	1,533	Wide temperature range, used in some nuclear reactors
Lead	327	1,744	High freezing temperature
Tin	232	2,600	Expensive, good heat transfer

Table 2.2 shows a list of some metal candidates. They all have higher maximum operating temperatures compared to solar salt and partly a lower freezing temperature. Additionally, they have very high heat transfer coefficients, thus high-solar fluxes on the receiver surface can be achieved while maintaining low surface temperatures. As a consequence, receiver apertures can be kept small and radiation losses are reduced. A good heat transfer is a major requirement for highly efficient receivers. Most metal candidates also have a low vapour pressure similar to solar salt, so the system can be non-pressurized.

Sodium has very good heat transfer characteristics; however, it is highly reactive with water and air in an exothermic reaction. Lead is not so reactive, but has a high-freezing point. The lead-bismuth eutectic and tin are promising metal candidates with a wide temperature range. However, due to their high price, there must be a heat transfer to a cheaper heat storage material.

2.5.2 Solar Tower with Particle Receiver and Storage

Another concept for high-temperature, high-flux receivers with the receiver working medium being the storage medium are particle receivers. Particles, e.g., doped (blackened) bauxite, with the size of about 0.5 to 1 mm are heated while being moved through the solar focus of a solar tower. These particles easily withstand temperatures in excess of 1000°C and they have no lower temperature limit as liquids. They can be stored easily in isolated containers under atmospheric pressure conditions.

The so-called solar falling particle receiver [11, 12] releases the particles above the solar focus and they heat while falling through it, see Figure 2.7 (left). The particle outlet temperature can be adjusted by adapting the particle

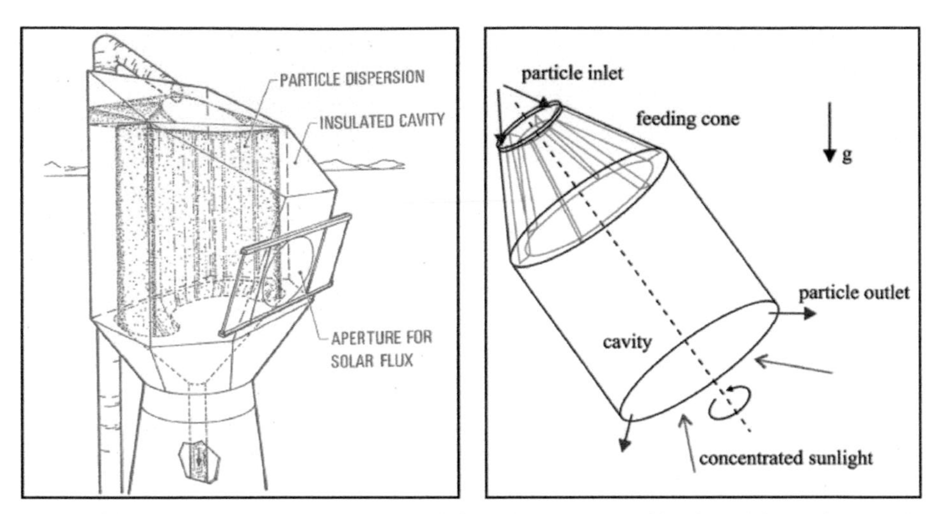

Figure 2.7 Falling particle receiver ([11], left) and rotating centrifugal particle receiver ([14], right).

mass flow rate. However, this may lead to highly transparent particle curtains for times with low solar insulation, resulting in low receiver efficiencies in part-load. Recirculation concepts [13] may overcome this non-ideal part-load behavior. Another concept to control the retention time of the particles in the solar focus is to use a rotating centrifugal receiver, in which centrifugal and resulting frictional forces control the particle movement, see Figure 2.7 (right) [14, 15].

The particle receiver belongs to the family of the direct absorption receivers (DARs). They couple the solar radiation directly into the storage medium, thus avoiding expensive high-temperature alloys for tubing and temperature gradients over the wall tubes, thus permitting high-solar fluxes. They have low sensitivity to peaks and transients in the solar radiation. Once filled the storage, the system allows a continuous operation of high-temperature processes for electricity generation or production of chemicals or solar fuels.

2.5.3 Molten Salt Large-Aperture Parabolic Trough

Higher concentrations and temperatures compared to the state of the art parabolic troughs (~5.8-m aperture width, ~80 suns, 390°C) can also be achieved by line-focusing technologies. Several companies did new collector designs and a scale-up of the collector aperture [16, 17] compared to the

Figure 2.8 Large-aperture troughs. Flabeg FE Ultimate Trough (Courtesy of Flabeg FE, left) and TSK Flagsol HelioTrough (Courtesy of TSK Flagsol, right).

standard parabolic trough (Figure 2.8). By using collectors of higher optical accuracy, the same receiver dimensions with 70mm absorber tubes can be used. A scale-up of factor 1.6, for example, increases the aperture width to 7.5 m, the solar concentration to ~105 suns, and allows higher temperatures of about 550°C without excessive radiation heat losses and hence good collector efficiencies. Due to the scale-up, the number of mirrors, receivers, pylons, sensors, joints, drives etc., and hence costs, are reduced.

The thermal oil-based heat transfer fluids have to be replaced in systems which operate above 390°C by others, e.g., silicone fluids, molten salt, or water/steam. Using molten salt as solar field working fluid and storage material has the advantage of reducing the number of heat transfer steps and complexity. Compared to thermal oil, molten salt has a better environmental footprint. The storage system can be smaller (~2.5 times) due to the fact that a higher temperature difference between cold and hot tank permits to store more energy per kg heat transfer fluid which reduces costs compared to a solar field operated on thermal oil. However, the operation of a molten-salt parabolic trough plant is more complex than a thermal oil-based plant, because part of the molten salt could freeze during failure or maintenance periods. A proper design for fail-safe drainage by gravity, for example, and emergency heat-tracing at selected points should solve these issues.

2.5.4 Combination of PV and modern CSP Plant

A further future concept which sounds attractive is the combination of a PV and a modern CSP plant as described above. Although, considering the energy system level only, it does not matter, if PV and CSP plant are separated or not. However, it should affect the project, commissioning, and operating costs.

The same project developer, constructor or operator may use synergies in the different project phases, such as solar resource assessment, site development, negotiations, bank loans, and permissions. They may share personnel and infrastructure like transformer stations, transmission lines, control rooms, or logistics.

The PV-CSP owner then can combine low-cost, intermittent PV with higher-value firm and flexible CSP and serve the market with a higher-value product and at larger volume (see red and orange areas in Figure 2.5). Market conditions will decide if this combination will be realized as physical plants side by side, or only as a virtual combination of two separate plants.

2.6 Summary

Concentrating Solar Power offers clean electricity and firm and flexible renewable capacity using a thermal storage. This is an additional value compared to intermittent electricity from wind and PV.

Future concepts of CSP will probably have high-solar concentrations for moderate high temperatures and highly efficient cycles using moderate temperatures [superheated steam, supercritical steam, and supercritical CO_2 (closed Brayton)]. Future systems certainly will include thermal storage to use the full potential of the technology. They will be reduced in complexity, e.g., using only one medium for receiver and storage system, or simpler heliostat and receiver layouts.

Examples for promising future concepts are central receiver systems with molten salt/molten metal receiver, central receiver systems with particle receiver, large-aperture parabolic troughs with molten salt and, combined CSP and PV plants to dispatch cheap and dispatchable electricity.

References

[1] National Renewable Energy Labs, NREL. (2015). *National Center for Photovoltaics, Chart 'Best research cell efficiencies'*, Rev. 08-06-2015. Available at: http://www.nrel.gov/ncpv/images/efficiency_chart.jpg

[2] International Energy Agency. (2014). *Technology Roadmap Solar Thermal Electricity*. Paris, France.

[3] Solar Power And Chemical Energy Systems, SolarPACES. *An implementing Agreement of the International Energy Agency*. Available at http://www.solarpaces.org/

[4] European Solar Thermal Electricity Association, ESTELA. Available at: http://www.estelasolar.org/techologies-plants/the-4-types-of-csp-electricity-technologies/

[5] National Renewable Energy Labs, NREL. *Concentrating Solar Power Projects*. Available at: http://www.nrel.gov/csp/solarpaces/

[6] International Renewable Energy Agency (IRENA). (2010). "Renewable energy technologies, Cost Analysis Series," in *Concentrating Solar Power,* Vol. 1, Power Sector, Issue 2/5, ed. Fichtner (Bann: International Renewable Energy Agency).

[7] Denholm, P., and Mehos, M. (2011). *Enabling Greater Penetration of Solar Power via the Use of CSP with Thermal Energy Storage*. Technical Report, NREL/TP-6A20-52978.

[8] Courtesy of T. Fichter, German Aerospace Center, DLR

[9] Litwin, R. Z. (2002). *Receiver System: Lessons Learned from Solar Two*. Sandia National Laboratory, Technical Report SAND2002–0084.

[10] Singer, C., Buck, R., Pitz-Paal, R., and Müller-Steinhagen, H. (2010). Assessment of solar power tower driven ultrasupercritical steam cycles applying tubular central receivers with varied heat transfer media. *J. Sol. Energy Eng.* 132. doi: 10.1115/1.4002137.

[11] Falcone, P., Noring, J., and Hruby, J. (1985). *Assessment of a Solid Particle Receiver for a High Temperature Solar Central Receiver System, SAND85–8208*. Albuquerque, NM: Sandia National Laboratories.

[12] Siegel, N., and Kolb, G. (2009). "Design and on-sun testing of a solid particle receiver prototype, ASME, ES2008," in *Proceedings of 2nd International Conference on Energy Sustainability*, Vol. 2, 329–334.

[13] Röger, M., Amsbeck, L., Gobereit, G., and Buck, R. (2011). Face-down solid particle receiver using recirculation. *J. Sol. Energy Eng.* 133. doi: 10.1115/1.4004269.

[14] Wu, W., Uhlig, R., Buck, R., and Pitz-Paal, R. (2015). Numerical simulation of a centrifugal particle receiver for high-temperature concentrating solar applications. *Numer. Heat Transf. A Appl.* 68, 133–149. doi: 10.1080/10407782.2014.977144.

[15] Wu, W., Trebing, D., Amsbeck, L., Buck, R., and Pitz-Paal, R. (2015). Prototype testing of a centrifugal particle receiver for high-temperature concentrating solar applications. *J. Solar Energy Eng.* 137. doi: 10.1115/1.4030657.

[16] Kötter, J., Decker, S., Detzler, R., Schäfer, J., Schmitz, M., and Herrmann, U. (2012). "Cost reduction of solar fields with heliotrough collector," in *Proceedings of SolarPACES 2012*, September 11–14, Marrakech, Morocco.

[17] Riffelmann, K., Richert, T., Nava, P., and Schweitzer, A. (2014). Ultimate trough: a significant step towards cost-competitive CSP. *Energy Proc.* 49 1831–1839. doi: 10.1016/j.egypro.2014.03.194.

3

Reinventing Geothermal Energy

Bodo von Düring

CLEAG Geothermal, Luzern, Switzerland

Abstract

Global market trends clearly indicate the rapid growth in renewable energy sources. Undoubtedly, a global demand for clean energy exists. However, there is a lack of economically feasible clean energy solutions, resulting in slow switchover rates.

CloZEd Loop Energy AG (CLEAG) offers an optimized variant of geothermal energy, which is characterized by a low CO_2 footprint, increased yield, lower emissions, and a new permanent CCU. CLEAG systems deliver a compact, decentralized, and independent energy supply. Although CLEAG power plants generate at base load levels of above 92%, their carbon footprint is close to zero emissions.

Our Croatian pilot project is supported by one of the world's largest funding programs for innovative low-carbon energy projects – NER300. The technology was developed in Switzerland in collaboration with the ecology center Langenbruck.

3.1 Classical Geothermal Energy

Deep geothermal energy uses water from great depths. Therefore, starting at 600 to 900 m depth, all dissolved gases are in liquid form. Once this water rises (pumped or artesian) to the surface, the gas expands in the ratio 1:3:1:5 and separates itself from the water. Today it is mostly released into the environment, without any further consideration.

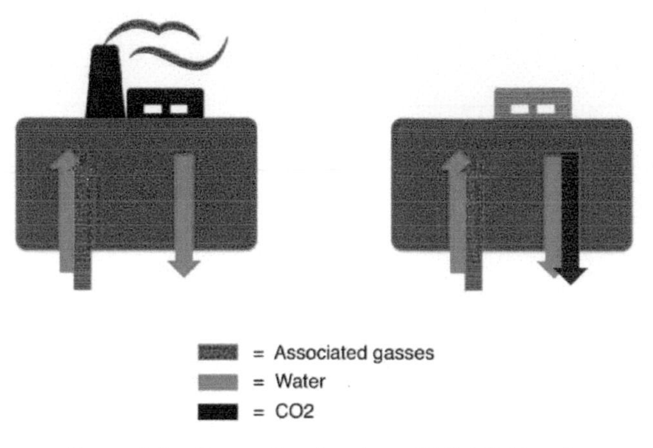

Figure 3.1 Conventional cycle and "CloZEd loop".

3.2 CloZEd Loop Energy AG

CloZEd Loop Energy AG endeavors to the principle "what you take out, you should also put back in," to keep the CLEAG cycle of geothermal fluids completely closed, and thereby creating the "CloZEd Loop." This allows the utilization of gases, which are flammable, in engines for the production of additional energy such as heat, cooling, and electrical power (Figure 3.1).

In contrast to conventional geothermal power plants, CLEAG excels by utilizing two sources of energy: hot water as well as the dissolved gases within this water. The gases are obtained solely from geothermal sources. Unlike traditional gas powered plants, the CO_2 is captured in the process, separated from the rest of the exhaust gases and safely brought back into the same aquifer. The gas extraction volumes of methane correspond exactly to those of the re-injection of CO_2. The advantages speak for themselves: it balances the subterranean surface and releases new methane from the sandstone within the aquifer (Taggert, 2010).

3.3 Optimization of the Efficiency

The power plants of CLEAG are designed so that they can be operated with minimal use of electricity, allowing further optimization of income. The today planned pilot plant consumes even in emergencies less than 20% own electricity. The running system requires only 10% own electricity and can thus achieve a high-net output.

Furthermore, it is important to mention that the plant will achieve an annual production of over 8,000 h (92% capacity rate) due to its concept of modular units. Thus, the plants have a base-load capability and achieve a very high income per annual production hour (MW/h).

3.4 How is this Optimization Achieved?

The following diagram describes the CLEAG concept in detail. CLEAG systems let the geothermal fluid run through a separator. In this separator, the gas separates from the water and is taken off the top. The water on the other hand, flows at the bottom of the separator and connects to a heat exchanger. After the heat exchanger, it flows at a cooler temperature to the re-injection pump. On the other hand, 42 MWt of water (>72°C) flow into the ORC Turbine, which are then converted into 3.3–3.6 MWe. This section of the CLEAG system represents the classical geothermal approach.

In the meantime, 31 MWt of gas are transferred to the combustion engine, which are converted into 14.4 MWe. The internal combustion engine operates with compressed air and the resulting exhaust gas contains nitrogen, carbon dioxide, and small amounts of nitrogen oxides. This exhaust is separated by an amine scrubber so that the CO_2 can be separated from the pure amine base using the waste heat from the engines. The CO_2 comes out with a temperature of >50°C and a pressure of 16 bar. In a shower tank the CO_2 is bound with the cooled water, which is then pumped at a pressure of less than 16 bar back into the aquifer. Due to this, the CLEAG system is able to be 5–6 times more efficient than conventional geothermal systems (Figure 3.2).

3.5 Performance Tuning

With a water-gas ratio (WGR) of 1:3 at an hourly capacity of 1,100 m^3/h and a content of combustible gases of about 97% the engines reach an electrical capacity of 13–14 MW. At the same time the ORC turbines from 97°C produce 3.7–4.9 MW of electricity. Moreover, due to the separation of gas from water, the ORC gains performance (Retrofitting). Both the heat from the water and the cooling water of the engines provide hot water for heating and cooling of 75 MW between 25–72°C. These figures demonstrate that the performance of CLEAG systems additionally provides almost three times the electricity compared with a traditional geothermal system. In particular, both the CO_2 and the methane is completely captured and pumped back into the same aquifer in the CCU process (Figure 3.3).

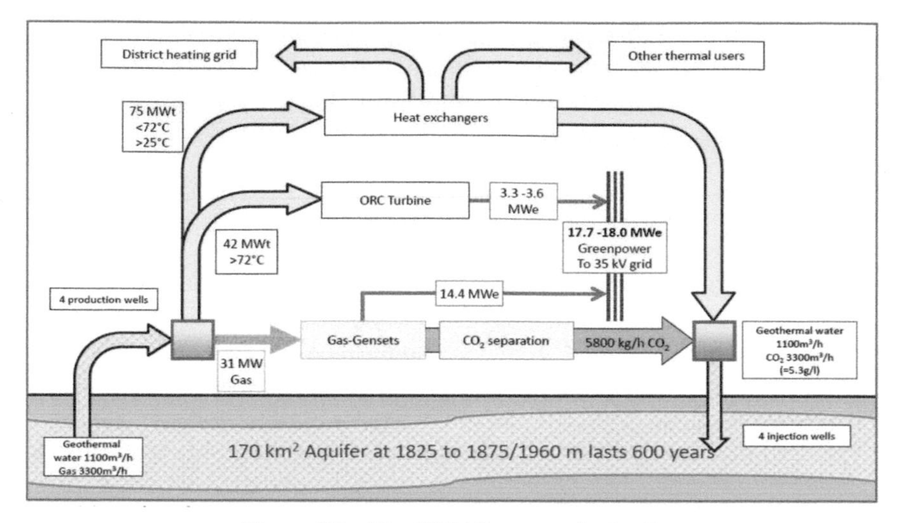

Figure 3.2 The CLEAG concept in detail.

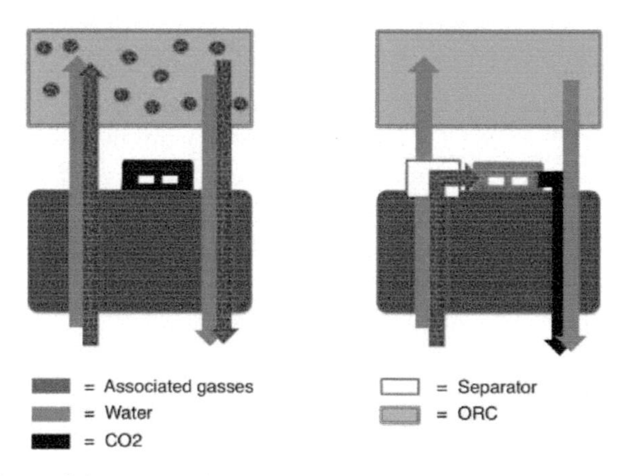

Figure 3.3 Retrofitting: Conventional ORC & CLEAG ORC.

3.6 CCU

CLEAG systems produce a permanently safe CCU. The injection in the 240 million years old aquifers is safe because even now the water contains CO_2. The amount of CO_2 will rise only minimally in 30 years and it is used as a chemical stimulant of the aquifer. Therefore, it can be expected that the amount of combustible gases will not decrease, because the re-injected CO_2, according to the research of Taggert (2010), releases new CH4 from the

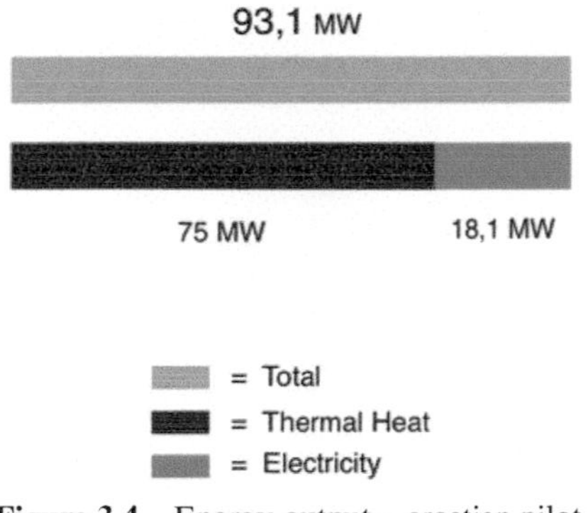

Figure 3.4 Energy output – croatian pilot.

sandstone. Furthermore, the re-injection serves to keep the aquifer balanced, consequently reducing the risk of earthquakes (Figure 3.4).

3.7 Levelized Cost

Levelized cost is defined as the total capital, fuel, and operating and maintenance costs associated with the plant over its lifetime divided by the estimated output in kWh over its lifetime. This diagram clearly shows that over the lifetime of the plant, geothermal power can be highly competitive with a variety of technologies, including natural gas (Figure 3.5).

3.8 Advantages and Disadvantages

The advantages of CLEAG are the low-carbon footprint and the increased investment income. The cost of investments increases only marginally compared to a conventional geothermal system. Moreover, aquifer gas resources are estimated twice as high as those of conventional gas and oil. Based on the CLEAG concept, nations can lower their dependency on oil, gas and electricity imports (Table 3.1).

A disadvantage is the relative restriction to an electrical capacity of less than 20 MW. Four production and four reinjection-drills are thereby required depending on the WGR. Nonetheless, this disadvantage translates into an advantage as it enables the usage of compact equipment that can be easily operated and maintained. Furthermore, it is ideal for establishing a decentralized energy supply. The heating/cooling can be used for housing, as well as stimulate local business growth.

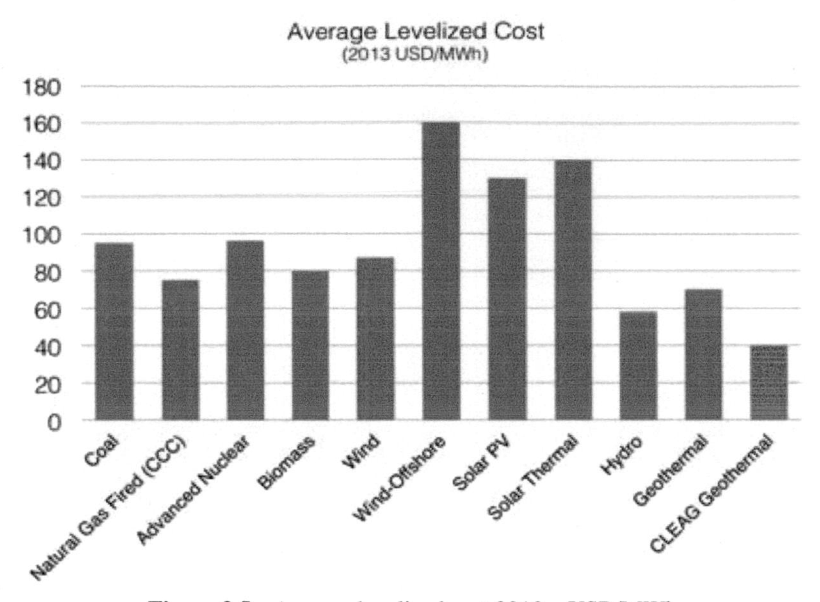

Figure 3.5 Average levelized cost 2013 – USD/MWh.

Table 3.1 Worldwide gas reserves & assumed resources (DIW Berlin, 2010)

(In billion m3)	Reserves	Resources	Total
Conventional Gas	188	239	427
Unconventional Gas	5	2720	2725
• Shale gas		456	456
• Other tight gas	3	210	213
• Coalbed methane	2	254	256
• **Aquifer gas**		**800**	**800**
• Gashydrat		1000	1000

3.9 Potential of the Pannonian Basin

The Pannonian basin is known as one of the very few geological formations in Europe ideally suitable for geothermal projects. It is one of the best explored territories in Europe. At the site chosen, water with a temperature of 100°C can already be found at 1,880 m depth.

Besides CLEAG's Croatian pilot project, the future potential for geothermal projects in the Pannonian Basin is tremendous. CLEAG envisions creating strong partnerships, with the same common goal of revitalizing the energy market by offering renewable, local and clean energy at affordable prices.

3.10 Use in Depleted Oil and Gas Fields

CLEAG's technology is especially interesting as application for depleted oil and gas fields. Wherever these fields have been filled with water, it usually has a content of hydrocarbons below the saturation point. This is an ideal starting point to build a CLEAG plant, which is able to continue the production of energy (heat, cooling, and electricity) as long as the hydrocarbon content of the water is high enough. This may be the case for years, because the re-injected CO_2 is helping to release new hydrocarbons from the formation.

3.11 Conclusion

CloZEd Loop Energy AG's approach is reinventing conventional geothermal energy. It is the response to uneconomic and environmentally harmful technologies, which are still being promoted today. By offering an optimized variant of geothermal energy, CLEAG necessitates the adaptation and improvement of current energy sources.

References

[1] DIW Berlin. (2010). Worldwide gas reserves and assumed resources.
[2] Taggert, I. (2010). "Extraction of dissolved methane in Brines by CO_2 injection: implications for sequestration."

4

From Waste to Energy: Development and Use of Renewable Energy in Sewage Treatment Facilities in Hong Kong

Dao Kwan-ming Keith

Project Division, Drainage Services Department, The Government of the Hong Kong, Special Administrative Region 44/F, Revenue Tower, 5 Gloucester Road, Wan Chai, Hong Kong, China

4.1 Introduction

Hong Kong, with an area of some 1,100 km^2, is a metropolitan city in Southeast Asia. With a population of about 7.3 million, Hong Kong generates more than 2.8 million cubic meters of sewage a day.

Drainage Services Department (DSD) of the Hong Kong Special Administrative Region (SAR) Government undertakes the responsibility for flood prevention as well as sewage collection, treatment, and disposal in Hong Kong. Currently, DSD operates more than 260 sewage and stormwater pumping stations as well as 70 sewage treatment facilities. With these, DSD treats more than 1,000 million cubic meters of sewage a year (Figures 4.1 and 4.2).

Over the years, with a vision "To provide world-class wastewater and stormwater drainage services enabling the sustainable development of Hong Kong", DSD has been, in the course of its operation, promoting energy efficiency via two approaches, namely (i) implementation of various energy saving measures and (ii) adoption of energy saving technologies to utilize various sources of renewable energy available in its sewage treatment works (STWs) and sewage/stormwater pumping stations. The sources of renewable energy include sewage sludge/biogas, wind and solar energy.

This article highlights the recent developments, challenges and outlook of utilization of renewable energy in DSD, focusing on the adoption of

Figure 4.1 Hong Kong.

Figure 4.2 The largest secondary sewage treatment works (STWs) in Hong Kong – Sha Tin STWs.

technologies such as combined heat and power (CHP), co-digestion of food waste with sewage sludge and photovoltaics (PV).

4.2 Biogas Generation and Combined Heat and Power Technologies

At present, around 93% of the population in Hong Kong are served by public sewers. The wastewater collected is treated to different levels, ranging from preliminary, primary, chemically enhanced primary, and secondary to tertiary treatment. Accordingly, about 900 tons of treated sewage sludge, or biosolids, are produced every day from DSD's STWs. Biosolids have a high calorific value, and typically, 1 kg of bone dry biosolids contains about 18,000 kJ. Its energy content is approximately 40% of that for gasoline. Biosolids are clearly a valuable renewable energy source.

DSD has adopted the anaerobic digestion process for treatment of biosolids at its four major regional secondary STWs. With the use of CHP/micro-turbine plants for co-generation, biogas produced in the anaerobic digesters as a by-product is largely transformed into useful electricity and heat energy. The electricity is used to power electrical equipment in STWs (such as sewage pumps and air blowers), whereas the recovered thermal energy is used to heat up and maintain the sludge inside the digesters at a temperature of about 35°C, for maintaining the performance of the digestion process.

The CHP plants installed in DSD's STWs are all running on reciprocating engines. From an engineering point of view, this kind of engines has the merits of quick starting and good part-load efficiencies.

Figure 4.3 indicates the normal operating process of a CHP system.

Biogas produced in sludge digesters is first transferred to and stored in biogas holders. After desulphurization and moisture removal, the purified biogas is fed at a constant pressure into the CHP plant. The energy stored in the biogas is then converted into thermal energy and mechanical energy which drives a synchronous generator for generating electricity to meet part of the power demand of the STW, with the heat recovered from the electricity generation process for maintaining the mesophilic digestion condition inside the sludge digester.

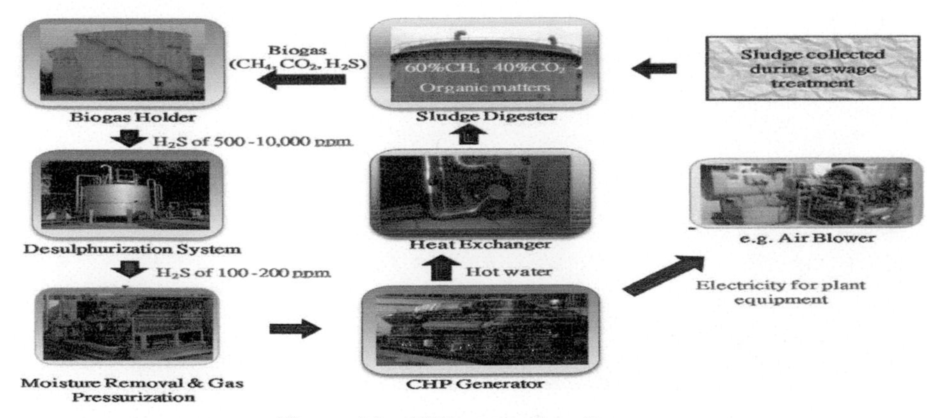

Figure 4.3 CHP system flow diagram.

The first CHP generator in DSD was commissioned in 2006. It has a capacity of 330 kW. Between 2006 and 2014, four more CHP plants and a micro-turbine generator were installed, bringing the total electricity generating capacity to 3,650 kW. Figure 4.4 presents the associated timeline of installation. All CHP plants in DSD are now operating in on-grid configuration (i.e., connected to operate in parallel with the power supply grid). Notably, the CHP plant in one of DSD's secondary STWs, with a capacity of 1.4 MW, is the largest high-voltage grid-connected generating unit operating in Hong Kong.

In 2014, the equivalent energy recovered from use of biogas in DSD's STWs was more than 28 million kWh, which is equivalent to the annual electricity consumption of some 3,100 four-member families and a reduction of emission of almost 20,000 tons of CO_2.

DSD's CHP systems did not get implemented without meeting some challenges. For example, if a CHP generator serves to power equipment that constitutes a large portion of electrical load (e.g., effluent pumps), the on–off

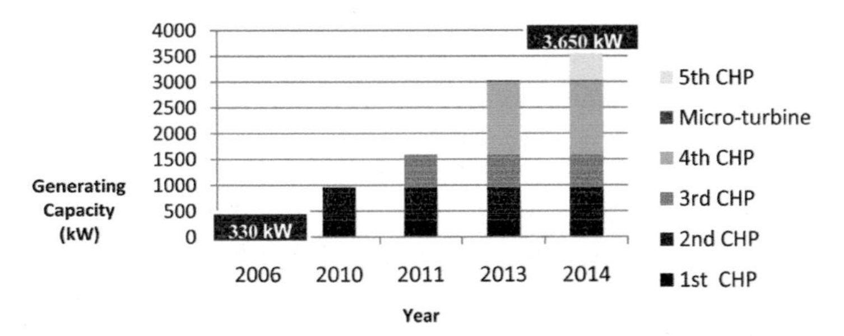

Figure 4.4 Timeline of installation of CHP plants and micro-turbine generator in DSD.

Figure 4.5 1.4 MW high-voltage CHP generator at Sha Tin sewage treatment works.

Figure 4.6 625 kW CHP plant at Tai Po sewage treatment works.

switching of the equipment sometimes causes nuisance tripping of the generator. This is because the generator is unable to reduce its operating frequency instantaneously and its protective device is thereby triggered. This accordingly causes temporary shutdown of the generator. A large pool of loading can help modulate the impact of switching of a single piece of equipment. As far as heating load is concerned, given that the sludge digesters in STWs require thermal energy for maintaining its optimum digestion temperature of some 35°C, the large amount of heat captured by CHP generators can perfectly be capitalized to heat up the sludge in winter days. However, in summer days, as the digesters do not need such a large amount of thermal energy, the heat generated by the CHP generators may not be able to be fully utilized and better utilization is now being explored, such as the use of tri-generation for meeting part of the cooling loads of the STWs.

To enhance the utilization of the biogas generated, which is on the rise, DSD is planning to install the second CHP generation system of a capacity of some 700 kW in one of its STWs.

4.3 Food Waste and Sewage Sludge (FS) Co-digestion

Currently, landfills are the ultimate disposal points of solid waste in Hong Kong. Without landfill expansion nor measures to drastically reduce the amount of solid waste, the capacity of the three existing landfills will be exhausted one by one by 2019[1] the latest.

[1]The Hong Kong SAR Government: Discussion Paper for the Legislative Council Panel on Environmental Affairs at 24 Feb 2014.

The average daily quantity of food waste being disposed in Hong Kong in 2012 was 3,337 tons[2], which accounts for 24.1% of its solid waste. Food waste, by virtue of its higher organic contents, has a higher specific energy value than municipal sewage solids.

In view of the high energy value of food waste, it is considered that if all the food waste were delivered to the sludge treatment facilities in STWs for co-digestion with sewage sludge to produce biogas, the amount of energy recovery could be remarkably enhanced and the burden to landfills can be alleviated. Figure 4.7 shows the co-digestion workflow.

Though FS co-digestion is a fairly well established technology which has been practiced overseas, there is a lack of reference in the application of the technology in Hong Kong. As such, a bench-scale laboratory test (Figure 4.8) on anaerobic co-digestion of food waste with sewage sludge is being conducted to ascertain the feasibility and performance of the technology in solid reduction and biogas production aspects at different operating conditions, such as (i) different solid retention time, (ii) different sewage sludge salinity (Note: some sludge in Hong Kong is saline because seawater is used for toilet flushing in many areas), (iii) different food waste/sludge ratio, and (iv) different types of food waste. Sewage sludge samples are collected from two secondary STWs while food waste samples are prepared based on the generic

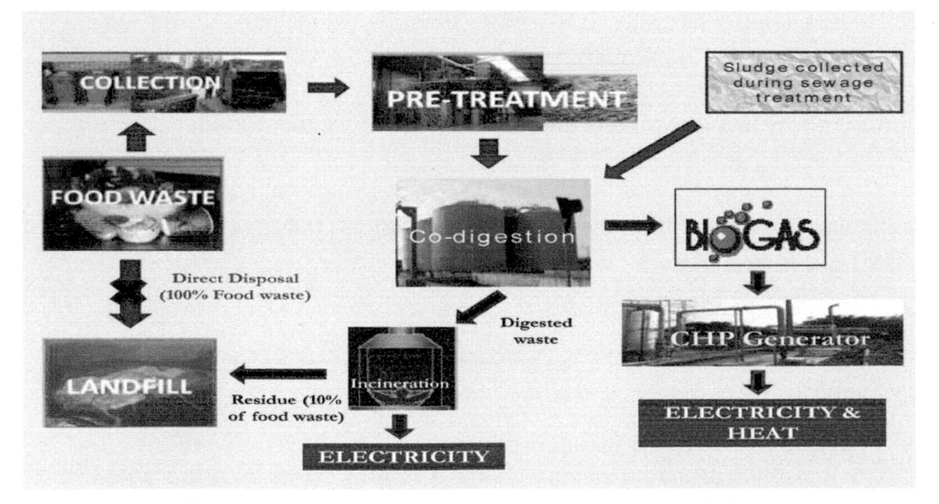

Figure 4.7 Co-digestion of food waste with sewage sludge.

[2]The Hong Kong SAR Government: Monitoring of Solid Waste in Hong Kong – Waste Statistics for 2012.

Figure 4.8 Bench-scale laboratory set-up for co-digestion.

food waste composition in Hong Kong. The solid content of feed sludge of co-digestion is controlled at 2–3% and the default co-digestion temperature is 35°C. It is expected that the results will provide essential information for taking further steps towards FS co-digestion, including an associated pilot plant trial.

The pilot plant trial will be conducted in one of DSD's secondary STWs with anaerobic digestion facilities to determine the feasibility, technical requirements, and performance of co-digestion (e.g., extent of reduction and dewatering of sludge as well as increase in biogas production). It is hoped that this can lead to formulation of a longer term implementation plan for territory-wide application of co-digestion of food waste with sewage sludge, and thereby bring about the benefits of better utilization of renewable energy, reduction of greenhouse gas emission and alleviation of the burden to landfills.

4.4 Photovoltaic (PV) Technologies

Over the years, DSD has endeavored to make full use of the open space in its STWs and pumping stations to capture solar energy for subsequent use as far as practicable.

Accordingly, DSD has installed various large-scale standalone and grid-connected PV systems for supplying electricity to equipment at its major facilities. The total capacity and annual electricity output of the PV panels were about 114 kW and 104,000 kWh per year, respectively, in 2014.

Besides, DSD has installed as a trial hybrid lamp posts, each equipped with a solar panel and a mini wind turbine, to capture both solar and wind energy in its STWs and pumping stations since 2011. Currently, the total capacity is about 49 kW (Figures 4.9 and 4.10).

Due to the lack of available open space for PV panel installation coupled with competing uses of space such as equipment installation in DSD's plants, space is always a constraint which limits large scale implementation of PV panels. Furthermore, possible glare caused by PV panels to neighboring community, remoteness of PV panels from electrical loads and their susceptibility to physical damage are amongst the constraints that have to be coped with.

At present, most of DSD's PV systems are installed at the roofs of its sewage treatment facilities and pumping stations. In order to identify more flexible locations for installation of PV systems and to save limited flat spaces for other uses, DSD is carrying out a pilot study on the use of vertical type

Figure 4.9 Rooftop PV panels.

Figure 4.10 Hybrid lamp posts.

building integrated photovoltaic (BIPV) system. Accordingly, a vertical type BIPV system is being installed at one of DSD's STWs to investigate the system's performance and efficiency with different solar cell materials (i.e., mono-crystallite silicon, poly-crystalline silicon and amorphous silicon). It is anticipated that the proposed vertical type BIPV system will not only allow better space utilization, but will also bring about an annual saving of some 1,750 kWh in electricity. This pilot study will provide useful reference for any future installation of such system in DSD's various facilities. Furthermore, capitalizing the opportunity of a recent update of a regional sewerage plan, which has made available a piece of land in Siu Ho Wan STW in the foreseeable future for PV panel installation, a PV system using poly-crystalline silicon will be installed there in phases in 2015 and 2016. With a proposed expansion of its output power from 850 to 1,100 kW peak, the system, when fully commissioned, would be the largest PV panel installation of the government as well as the largest in Hong Kong.

4.5 Conclusion

Population growth, economic development and rising public aspiration for better quality of life and sustainable development have raised the demand for an environmentally friendly and energy efficient sewage treatment systems in Hong Kong.

Biogas from sewage sludge, which used to be considered as waste, and solar radiation are major sources of renewable energy that can be found in DSD's sewage treatment facilities during the day-to-day operation. The planned use of CHP plants is considered to be one of the most practical and effective means for turning biogas into useful energy. Up to now, the total installed electricity generating capacity of renewable energy systems in DSD is around 8.2 MW and the electrical and thermal energy saved by the systems was about 29 million kWh in 2014.

Looking ahead, while a study is underway to install the 6th CHP system in DSD's STWs to optimize biogas utilization, other studies are being carried out to boost biogas generation from sewage sludge by adopting new technologies/processes. The new technologies/processes include anaerobic digestion of chemically enhanced primary treatment sludge and ultrasonic pre-treatment of sludge to enhance anaerobic sludge digestion. Besides, the feasibility of employment of fuel cells to utilize biogas to co-generate electricity and heat is also being examined through laboratory tests. Separately, the operating requirements and performance of FS co-digestion, which can effectively turn food waste into useful energy, increase the yield of biogas and reduce the burden to landfills, are being studied in a bid for making contributions to the Hong Kong's environment.

To further promote the use of solar energy, the largest PV system in Hong Kong is planned to be constructed in one of DSD's STWs.

DSD will continue to keep abreast of the latest development in energy saving technologies and renewable energy utilization, and to strive to reduce energy consumption as well as greenhouse gas emission in the course of its operation. It is envisaged that new technologies emerging in the not too far distant future will help further enhance the performance of the systems that utilize renewable energy, whereby providing even more contributions to sustainable development.

References

[1] COGEN Europe. (2001). *A guide to cogeneration.* Brussels: The European Association for the Promotion of Cogeneration.

[2] Legislative Council Panel on Environmental Affairs. (2014). *The Hong Kong SAR Government: Discussion Paper for the Legislative Council Panel on Environmental Affairs.*

[3] Drainage Services Department. (2014). *The Hong Kong SAR Government: DSD Sustainability Report 2013–2014.*

5

Cacao and Coffee Roasting Using Concentrated Solar Energy: A Technology for Small Farmers in Rural Areas of Peru

François Veynandt, Juan Pablo Perez panduro and Miguel Hadzich

Concentrating Solar Technologies and Building's Energy Efficiency,
Ecole des Mines d'Albi Campus Jarland, 81013 ALBI Cedex 9, France

Abstract

Solar food processing is gaining interest for income generation. The solar roaster from GRUPO PUCP is designed for rural areas. It consists in a horizontal rotating drum, opened at one end to collect solar radiation from a Scheffler concentrator of 2.7 or 8 m^2. The experimental campaigns brought valuable knowledge on the system's operation. Considering the promising results obtained, two initiatives help spread the technology: Acacao Solar is a pilot plant for cacao paste production and Compadre is a startup which enable small farmers to roast themselves their coffee with the sun.

Keywords: concentrating solar thermal energy, solar cooking/food processing, cacao/coffee roasting, low-tech, open-source, rural sector, Peru.

5.1 About Solar Roasting

There are few documented experiences on solar roasting. Solar Roast Coffee, USA, started roasting coffee using concentrated solar thermal energy. They built roasters of increasing scale and claimed a high-quality product [1]. ChocoBiciSolar, in Mexico, produces chocolate using solar energy and bicycle powered machines. Started in 2004 as ChocoSol, several hundreds of kilograms of cacao have been roasted in the first year, with a technology based on

Figure 5.1 Solar Roast Coffee, USA (left) and ChocoSolar, Mexico (right).

the solar concentrator now diffused by GoSol [2]. In Namibia, Nailoke Solar House roasts peanuts in a solar box cooker. The roast is more homogeneous than with the traditional coal or wood fire. In Burkina Faso, ULOG Freiburg and ISOMET developed a solar powered Shea butter production unit. The roasting step is performed in a pot with constant manual stirring using a paddle ensuring an even roast of the Shea nuts [3] (Figure 5.1).

5.2 An Opportunity for Social and Environmental Benefits

Cacao and coffee are important agricultural products in Peru. They are mainly grown at small scale by over 250 thousand families managing less than 5 ha. Their cacao and coffee are mostly bought as dry raw beans at low and unstable prices, making this activity barely economically viable. Enabling small-scale growers to process their cacao and coffee beans can improve their quality of life with higher incomes and better perceived value of their contribution to the society.

The solar roasting technology presented here has been developed by the *GRUPO PUCP*. The *Grupo de Apoyo al Sector Rural* is the support group to the rural sector of the *Pontificia Universidad Catolica del Peru* (PUCP). The *GRUPO PUCP* develops since more than 20 years appropriate technologies for the rural areas of Peru [4]. The solar roaster is designed for the rural sector, using mainly accessible material and locally available know-how. The machine obtained is robust, low tech, low cost, easy to maintain and to handle locally. The aim is to achieve the highest quality with a technology both environmentally and socially responsible. This applied research has been funded by the PUCP in 2012 and FINCyT (Fondo para la Innovacion, la Ciencia y la Tecnologia) in 2014–2015.

5.3 Description of the Solar Roasting Technology

The technology presented here is the result of the experimental study led in Huyro, in the heart of a coffee and cacao producing region. Photos on Figure 5.2 illustrate the system in the test field of the *"Granja Ecologica"*, Huyro. The solar radiation there is good, with an average annual Direct Normal Irradiation (DNI) of 2040 kWh/(m^2.an) (Lat.: 13.01°; Long.: –72.56°) [5]. The harvest season (April to September) is especially sunny, which encourages the use of solar energy for cacao and coffee processing.

The technology developed is based on the open-source Scheffler solar concentrator [6, 7]. A specific horizontal rotating drum hosts the beans and it is placed at the focal point.

The shape of Scheffler concentrator is a lateral section of parabola. It offers a fixed focal point during the day and the year, which makes it suitable for building integration. Tracking is easy on a single mechanically balanced axis. A seasonal adjustment of the parabola's curvature enables to adapt to the variations of the sun altitude throughout the year. The 2.7 m^2-mirror-surface version can deliver 1 kW of useful thermal power (at 100°C). It has a peak temperature of about 300–400°C. A larger version of 8 m^2 has also been built and tested for cacao roasting.

The drum is opened at one end, to let enter the solar light from the concentrator. The beans of cacao or of coffee are heated directly by concentrated solar light. No glass cover is used to tap the aperture. This ensures the evacuation of the moisture, the gaseous products and the "silver" skin through this opening.

Figure 5.2 Solar roasting technology with the concentrator of 2.7 m^2 (left) and 8 m^2 (right).

For the 2.7-m^2 concentrator, the drum used is 28 cm long and has a diameter of 25 cm. For the 8 m^2 one, the drum has a length of 48 cm and a diameter of 34 cm. The drum is made of black enameled steel to absorb light on the inside wall. Thermal insulation is placed around the drum. This ensures good thermal efficiency and improves the roasting process.

5.4 Sharing the Operational Experience

The experimental study led to an optimized configuration with high quality: coffee has been rated premium (over 80 points/100) according to the SCAA's Coffee Classification and cacao also gave promising results.

Depending on the direct solar radiation, the 2.7-m^2 Scheffler concentrator should operate with 1–2 kg of coffee per batch, to respect the optimal roast time of 15 to 20 min. Cacao is less sensitive to longer roasting time, so 2 kg can be roasted in the 2.7 m^2 concentrator and 6 kg in the 8-m^2 concentrator. A backward inclination of 15–20° is sufficient to make the maximum quantity fit in the drum. A rotational speed of 20 rpm ensures an even roast.

The roasting procedure consists in: (i) *Setting the Scheffler concentrator* by rotating it on its axis to reflect the light toward the drum. The position of the reflector is adjusted manually every 5 min to track the sun. (ii) *Preheating the empty drum* to at least 120°C for cacao or 150°C for coffee. (iii) *Feeding the cacao or the coffee beans* in the rotating drum. (iv) *Roasting* by heating the beans up to 120–160°C for cacao beans or 200–230°C for coffee beans. Temperature evolution is tracked by an infrared sensor or thermocouple. (v) *Extracting the beans* from the drum. (vi) For cacao: *Husking and grinding* to obtain pure cacao paste. For coffee: *Cool down* quickly to stop the roasting of coffee beans at the exact point desired.

5.5 Temperature Profile During Roasting

On Figure 5.3 (left), 3 kg of cacao beans have been roasted under constant solar radiation of 830 W/m^2. The drum is preheated to 160°C in 3.5 min. After 4.5 min, the cacao is poured into the drum. The beans temperature increases with an initial heating rate of 8.5°C/min, slowing down to 1.4°C/min. Indeed, the higher the temperature, the stronger the thermal losses: conduction through the drum's walls, convection with the atmosphere and infrared radiation with the environment. Roasting ends after 33 min, when cacao's temperature reaches 140°C.

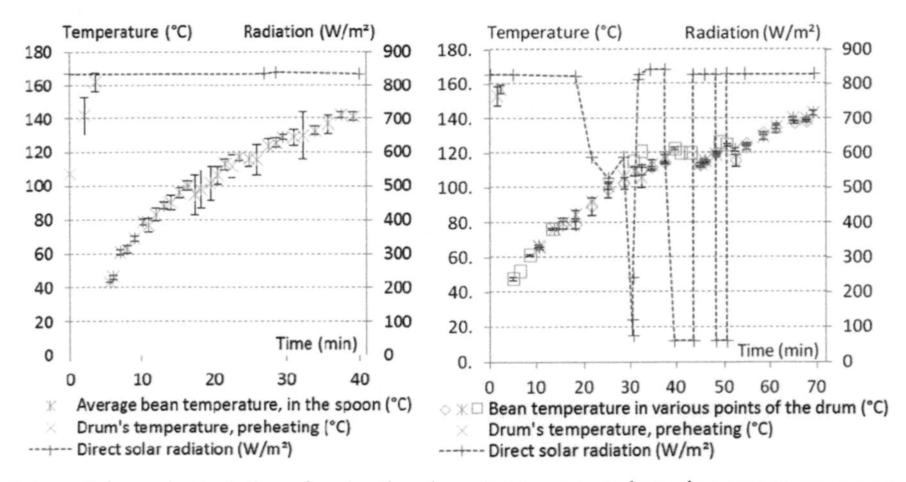

Figure 5.3 *Left:* Evolution of preheating drum temperature and roasting cacao temperature, good direct solar radiation. *Right:* Evolution of cacao temperature, effect of small clouds.

Figure 5.3 (right) shows roasting while having short term clouds (radiation <100 W/m^2). The inertia of the drum and the inertia of the 6 kg of cacao maintain the system's temperature during several minutes: a 2°C drop in 6 min is observed during the second cloud.

Coffee roasts in a similar way. Figure 5.4 shows the temperature evolution of coffee during roasting and during the following cool down, after taking out the coffee beans from the drum. The internal drum wall temperature is also monitored. The coffee heating rate slows down in the first 7 min.

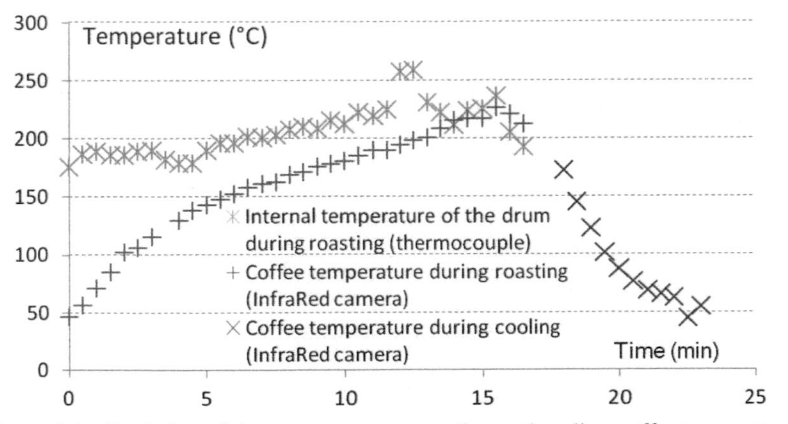

Figure 5.4 Evolution of drum temperature, roasting and cooling coffee temperature.

When reaching 160°C, because of its changing color from light green to dark brown, the absorptivity of coffee increases and the heating rate is maintained constant up to the final temperature of 230°C. In this experiment, 750 g have been roasted in 17 min under a sun of 830 W/m^2. The cooling down is performed by simple manual forced convection bringing the beans temperature to 50°C in 5 min.

5.6 Diffusion of the Technology

Given the good results of the experimentations, the technology is being promoted. Two initiatives have been started, as described in the following: the demonstration plant for cacao transformation and a startup to make the solar coffee roaster accessible to the farmers.

5.7 Demonstration Plant of Pure Cacao Paste

To demonstrate the opportunity of solar roasting to the local producers, a pilot plant *Acacao Solar* has been built and is freely available in Huyro, region of Cusco. Figure 5.5 illustrates the plant, which is equipped of two 8-m^2 Scheffler concentrators, for a daily production capacity of 100 kg of pure cacao paste.

5.8 Startup Compadre

Compadre empowers the small-scale farmers to roast themselves their coffee. For that, the startup company brings the technology and helps them to commercialize their product. This promotes a different distribution of the added value. Another example is in the wine production, where many producers process their grapes themselves. In countries like Peru, it can help decrease the rural exodus by bringing interesting economical activities in the rural areas.

Compadre is in its first year of operation, learning a lot on how to operate successfully. It has been launched with the support of *UTEC Ventures* and *Startup Peru*. The team of the *GRUPO PUCP* is happy to see that its research on this appropriate solar roaster brings effectively an interesting alternative in rural economy.

The action plan is to install a production facility with the technology, tools, and material needed for coffee roasting and packaging the final product. A training program is provided to the farmers in order to use the solar roaster. An online platform creates a short supply chain: the consumer can access the

Figure 5.5 Views of *Acacao Solar*, the solar fabric transforming raw cacao beans into cacao paste.

farmer's profile and order online and the farmer can get feedback from its customers. *Compadre* envisions the implementation of further applications as described in the following perspectives (Figure 5.6).

5.9 Perspectives

This solar roaster can enhance economic situation in rural areas, especially considering its versatility: roasting of coffee, cacao, broad beans, and production of pop-corn. Additionally, the Scheffler concentrator itself opens the opportunity for other applications like: cooking, milk pasteurization, essential oil extraction, distillation, steam production. This encourages further optimization and promotion of the technology, as well as other solutions using renewable energy and appropriate technologies. This will contribute to empower people in rural areas and help them keep living on their land and with their culture.

Figure 5.6 *Compadre*'s solar roasters and the coffee produced in Huyro and Satipo, Peru.

References

[1] SOLARROASTCOFFEE. (2014). *Solar roast coffee. artisan roasting with solar power*. Available at: http://www.solarroast.com/src_about

[2] Cocina solar mexico. (2015). *Chocobicisolar*. Available at: http://www.cocinasolarmexico.com.mx/crbst_7.html

[3] Solar food processing network: Projects. Available at: http://www.solarfood.org/projects

[4] GRUPO PUCP. *Grupo de Apoyo al Sector Rural*, Pontificia Universidad Catolica del Peru. Available at: http://gruporural.pucp.edu.pe/

[5] NASA. (2015). *Nasa surface meteorology and solar energy – available tables*. Available at: https://eosweb.larc.nasa.gov/cgi-bin/sse/grid.cgi?&num=108077&lat=-13.01&submit=Submit&hgt=100&veg=17&sitelev=&email=skip@larc.nasa.gov&p=grid_id&p=avg_dnr&step=2&lon=-72.56

[6] Munir, A., Hensel, O., and Scheffer, W. (2010). Design principle and calculations of a scheffer fixed focus concentrator for medium temperature applications. *Solar Energy* 84, 1490–1502. doi: 10.1016/j.solener.2010.05.011.

[7] Solare Brücke. (2015). *The scheffer reflector*. Available at: http://www.solare-bruecke.org

6

Socioeconomic, Environmental, and Social Impacts of a Concentrated Solar Power Energy Project in Northern Chile

Irene Rodríguez[1,*], Natalia Caldés[1], Alberto Garrido[2], Cristina De La Rúa[1] and Yolanda Lechón[1]

[1]Energy Department, Energy Systems Analysis Unit, CIEMAT, Av. Complutense 40, E-28040 Madrid, Spain
[2]CEIGRAM, Research Centre for the Management of Agricultural and Environmental Risks, Av. Complutense s/n, E-28040 Madrid, Spain
*Corresponding Author: irene.rodriguez@ciemat.es

Abstract

Concentrated Solar Power deployment could potentially play an important role in the sustainable development strategy of Chile, the country with the highest solar potential in the world. In this regards, besides electricity generation costs, it is also important to assess the socioeconomic, environmental and social implications of any energy investment project. In order to shed some light to this issue, this work contributes to enlarge the existing body of knowledge by conducting a sustainability assessment of the installation, operation, and maintenance of a 110-MW Concentrated Solar Power tower plant in Chile. Using an Input–Output methodology and based on plants costs data, this work estimates the socioeconomic and environmental direct and indirect effects of the project in terms of economic activity, job creation, energy consumption, and CO_2 emissions. Additionally, using the Social Hotspot Database, a preliminary social risk analysis in those economic sectors most stimulated by the project in terms of employment is performed. Assuming domestic provision of all goods and services, results show that the associated total socioeconomic impacts during the whole lifetime of the plant would amount to 3,124 million US$, a multiplier effect of 2.2 and a ratio of indirect per direct job creation of 1.21. Additionally, results also show that direct and indirect economic activities required by the project would generate 64.36 g CO_2 per kWh. Finally, the social assessment indicates the existence of a high unemployment risk in those sectors most stimulated, therefore, the project could decrease these unemployment risks.

Keywords: input–output analysis, sustainability impact assessment, concentrated solar power, Chile.

6.1 Introduction

Over history, energy has been an essential driver for the development of civilizations (Kammen and Dove, 1997). At the same time, the energy sector is one of the largest consumers of natural resources and responsible for greenhouse gas (GHG) emissions. In this sense, a sustainable approach which accounts for the harmony between economy, society and environment must be put in place (World Commission on Environment and Development, 1987).

One way to reduce GHG emissions worldwide is to progressively substitute fossil fuel technologies by renewable ones. The country with the highest solar energy potential is Chile, in particular the Atacama Desert in the North of the country (Seminario Iberoamericano de Energías Renovables, 2009). The need to promote a wide portfolio of renewable energy technologies in Chile is critical due to the seasonality of its large hydropower energy and also its large energy dependency from fossil fuel, mostly imported (Larraín and Escobar, 2012). The Chilean Energy Commission notices that 82% of the electricity of the Northern Chile Electric System (SING) comes from coal resources (Comisión Nacional de Energía, 2012).

For this reason, the Chilean government introduced in 2008 the Non-Conventional Renewable Energy Law 20.257 (excluding hydropower), which consists on increasing renewable energy production to 10% of the total energy mix in 2024 (Ministerio de Energía, 2008). This law was renewed in 2013 (Law 20/25) by increasing the renewable energy target to 20% in 2025 (Ministerio de Energía, 2013). Moreover, in 2014 the "Energy 2050" program was launched, which constructs a long-term vision in the Chilean energy system to 2050. It consists of four steps (Ministerio de Energía, 2014): energy agenda (end of 2014), roadmap to 2050 (first semester 2015), long term energy policy (end of 2015) and dissemination (first semester 2016).

In Chile, due to its outstanding solar resource, Concentrated Solar Power (CSP) is one of the most promising renewable energy technologies. Compared to other technologies, the main advantage of CSP is its storage capacity

(CSPToday, 2012). The capacity factor of CSP plants increases with energy storage systems, ranging from around 20% without storage system, 40% with 6 h of storage and 60% with more than 12 h of storage (Hoffschmidt et al., 2012).

At the same time, compared to other renewable energy technologies, its main disadvantage is its higher cost (Avila-Marin et al., 2013). However, recent research indicates that its Levelized Cost of Energy (LCOE) will experience a considerable decrease over the period 2010–2030, and a slower but continuous rate between 2030 and 2050 (Hernández-Moro and Martínez-Duart, 2013). It is expected that the main LCOE reduction will come from the solar field (15–16%) by improving economy of scale; low values of turbine power output (33–78 MWe) and very high thermal energy storage systems (14–16 h; Avila-Marin et al., 2013). According to various studies, as a result of these improvements, a 25–30% reduction of the LCOE is expected. Also, it is important to consider that LCOE also depends on the plant location, decreasing in countries with the highest solar irradiation (Hernández-Moro and Martínez-Duart, 2013). The most suitable Chilean region in terms of solar irradiation is Region II Antofagasta, within the Atacama Desert. Antofagasta has great energy demands mainly from the mining sector which, in 2012, represented 61.1% of its regional Gross Domestic Product (GDP) (Banco Central de Chile, 2012).

Thus, given the large energy consumption of the mining sector, the huge solar potential and the expected cost reduction of the technology, CSP is likely to play a key role in the energy mix and contribute to the decarbonization efforts of the country. In this sense, this work conducts a sustainability assessment of the flagship CSP project in Antofagasta "Cerro Dominador" applying the Input–Output (I–O) framework as well as the Social Hotspot Database (SHDB).

6.2 Methodology

Input–output methodology analyzes the response of the economic sectors due to an increase of the demand of goods and services generated by a project (Ten Raa, 2006). This methodology was first developed Leontief (1936). Nowadays, it is a robust tool widely used, and the energy sector is not an exception (Linares et al., 1996).

I–O analysis is based on the I–O tables (IOT), which displays the intersectorial relations among the economic sectors of a country. From the IOT, it is possible to obtain the technical coefficients, which indicate the consumption that one sector requires from another sector to produce one single unit monetary unit (Equation 6.1) (De la Rua Lope, 2009):

$$aij = xij/Xj. \tag{6.1}$$

where: *xij* is the amount of product that the economic sector *j* requires from the economic sector *i* to generate its final production *X;* and *aij* is the amount of product that the economic sector *j* requires from the economic sector *i* to produce one unit of product of *j*.

Additionally, the IOT display the added value of each economic sector and the final consumption of private and public sectors (Tarancón, 2003). Equation 6.2 shows the final production calculation, which accounts for both direct and indirect effects in the economy. Direct effects refer to the ones produced by the increase in the demand of those economic sectors that directly provide the goods and services required to the construction, operation and maintenance of the plant; and indirect effects are the ones produced by the effect that this new investment has on new flows of purchases and sales among economic sectors (Caldés et al., 2009).

$$X = (I - A)^{-1}Y \tag{6.2}$$

where: X is the final production; A is the technical coefficient matrix; Y is the final consumption demand; and $(I - A)^{-1}$ is the Leontief inverse matrix, which quantifies the direct and indirect requirements to satisfy a certain final demand.

Due to the interdependency among economic sectors, the development of any project involves a widespread stimulation of various sectors in the economy. Such effect is the so called multiplier effect and indicates how much the total income of a country increases for every monetary unit invested in a project. The multiplier effect is the ratio between total effects (direct and indirect effects) and direct effects (Holland and Cooke 1992).

Moreover, the methodology allows estimating the induced effects which account for the economic effects generated by the expenditures of the workers involved in the project (Caldés and Lechón , 2010).

An extension of this methodology consists on assessing other type of effects – such as job creation and CO_2 emissions – per unit of output for each economic sector. The estimation of such effects is shown in Equation 6.3 (Caldés and Lechón, 2010):

$$\Delta X \times Zi = (I - A)^{-1} * \Delta Y \times Zi, \tag{6.3}$$

where: ΔX is the total (direct and indirect) increase in goods and services; $(I - A)^{-1}$ is the Leontief inverse matrix; ΔY is the final demand and Zi is the environmental or socioeconomic vector, which indicates the employment, emissions, energy consumption, etc., per unit of production for each economic sector included in the IOT.

One of the advantages of the methodology is that the IOT are usually available from the National Statistics databases. However, it also faces some limitations: the production capacity is assumed to be unlimited; it does not account for the possibility of storage and, finally, all informal transactions in the economy are not accounted for. Also, IOT are only published and updated every few years, which prevents the analysis from considering some relevant changes in the economy (e.g., technology improvements; De la Rua Lope, 2009).

With respect to social impacts, the Social Hotspot Database (SHDB) is used to identify the main social risks in those sectors more affected by the project in terms of employment. The SHDB differentiates social risks into five different impact categories[1] in a specific economic sector in a specific country. Moreover, the SHDB allows assessing the risk of the themes and subthemes within a social impact category. The different degrees of the social risk are classified as "very high", "high", "medium", and "low". Finally, the so called "Social Hotspot Index" (SHI) compares, in a quantitative way, the different social risks among sectors and countries. The index formation is based on the transformation of qualitative data into quantitative data of the risk of each social theme. The final SHI of each impact category results from the addition of all social risks of all the social themes within an impact category (Equation 6.4). A SHI close to a hundred value means that there exist high or very high social risks within an impact category (GreenDelta, 2013).

$$\mathrm{SHI}_{\mathrm{cat}} = \Sigma(R_{\mathrm{avg}} \times W_T / \Sigma(R_{\mathrm{max}} \times W_T), \quad (6.4)$$

where: $\mathrm{SHI}_{\mathrm{cat}}$ is the index of the social risk of each impact category; T is the social theme; n is the number of themes within an impact category; R_{avg} is the average risk of each social theme and R_{max} is the maximum risk of each theme and W_T is the assigned weight to each social theme.

The principal advantage of the SHDB is that not only the most common socioeconomic impacts (like employment) are considered but also other social

[1]The categories are "labor rights and decent work", "healthy and safety", "human rights", "governance", and "community infrastructure".

impacts not usually accounted for (human rights, labor rights, cultural aspects, etc.; GreenDelta, 2013).

6.3 Data and Assumptions

The principal assumptions considered in this work are (i) domestic manufacturing of all project components (no imports assumption); (ii) 1 year of plant construction; (iii) 30 years plant lifetime in the operation and maintenance (O&M) phase; (iv) non-use of additional energy fuel (e.g., natural gas); and (v) the electricity consumption of the plant comes from the Chilean electricity system (without self-consumption).

Table 6.1 shows the principal characteristics of Cerro Dominador project based on (CSP World database) and (ABENGOA, 2013).

6.3.1 Direct, Indirect, and Induced Economic Effects; and Employment Vector Data Source

The direct and indirect economic effects have been calculated based on the Chilean IOT and investment and O&M cost data from other plants.

The Chilean IOT has been obtained from the Organization for Economic Co-operation and Development (OECD) statistics database. The IOT reference year is 2003, it contains 37 economic sectors and the monetary unit is million Chilean pesos which, for this work, have been converted to US million dollars (US$M).

Table 6.1 Principal characteristics of Cerro Dominador plant

Owner	Abengoa
Location	Maria Elena, Antofagasta
Status	Under construction
Operation start date	June 1018
Power	110 MW
On-peak capacity factor	94.5%
Energy in a year	910,602 MWh
Energy generated in the all life	27,318 GWh
Land area	1,400 ha
Technology	Central receiver
N° heliostats	10.600
Storage hours	17.5
Type of storage	Sodium and potassium molten salts
Type of cooling system	Dry
Investment cost	1,300 millionUS$ (US$M)
Use of electricity	Supply to SING

The investment costs have been estimated based on cost data of a 17-MW Spanish plant (Caldés et al., 2009) and have been extrapolated to a 100-MW plant with 15 storage hours (Fichtner, 2010). The case study investment cost breakdown is described below: 37% solar field, 14% tower, 6% storage system, 12% power block, 1% land, 8% engineering and insurances, 6% construction, 6% financial costs, 5% balance of plant, and 5% other expenses.

The total investment cost assumed is 1,338 US$M, which is similar to the one in the Environmental Impact Statement (1,300 US$M) (ABENGOA, 2013). A cost reduction of 12% has been assumed by 2015 according to the International Renewable Energy Agency (IRENA) (Hoffschmidt et al., 2012).

O&M cost estimations have been decided based on an extensive literature review. The annual O&M cost of a 100-MW plant with 9 h of storage is estimated at 6,500,000 $/year (Kolb et al., 2011). Based on this reference and an O&M cost breakdown (Fichtner, 2010), it has been assumed that the O&M cost without personnel costs for this case study would amount 4,877,704 $/year. The discount rate considered for the current conversion of future prices is 5% (Caldés et al., 2009). The O&M breakdown in this case study is described below based on (Fichtner, 2010): 31% solar field and storage system; 20% financial costs; 25% personnel costs; 14% power block; and 10% variable costs.

Table 6.2 displays all the plant costs assigned to the different economic sectors of the Chilean IOT in both investment and O&M phases based on the Spanish plant (Caldés et al., 2009).

For the assessment of the induced effects, personnel cost data, their propensity to consume and the distribution of the households' expenditures have been considered. Personnel cost data assumption is 1.622.296 $/year based on (Fichtner, 2010) and (Kolb et al., 2011). The propensity to consume considered for Chile is 0.67 (De Gregorio, 1998). The Chilean household expenditures distribution in % across economic sectors in 2003 can be consulted in (Chile Central Bank, 209).

Regarding employment, Table 6.3 shows the total employment data by sectors in Chile in 2003 from the Chilean National Statistics Institute (INE) (Instituto Nacional de Estadísitica, 2003).

In order to overcome the data gap between the 9 economic sectors for which the Chilean INE provides employment data and the 37 sectors of the IOT, the employment vector of Brazil from the World IO Database (WIOD) has been considered; except for the agriculture, mining and the construction sector, which do not need any desegregation.

Table 6.2 Costs assigned to each economic sector of the Chilean IOT in both investment and O&M phases

	Investment Phase		O&M Phase	
IOT Sector	Percentage to Sectors (%)	1,000$	Percentage to Sectors (%)	$/year
8-Chemicals	6.6	99,694	4	195,108
9-Plastics	1.5	22,395	1	48,777
10-Non-metallic minerals	27.7	421,495	6	292,662
11-Basic metals	4.3	65,885	–	–
12-Fabricated metals	19.5	296,698	–	–
13-Machinery and equipments	9.1	138,533	1	48,777
14-Office and computer machinery	–	–	3	146,331
15-Electrical machinery	11.0	167,121	2	97,554
21-Electricity, gas, and water	0.1	1,512	10	487,770
22-Construction	6.7%	101,579		
26-Post and telecommunications	–	–	15	731,656
27-Finance and insurance	5.4	82,041	25	1,219,426
28-Real estate activities	0.1	1,787	–	–
29-Machinery and equipment renting	0.1	1,100	4%	195,108
30-Computer activities	0.1	962	4	195,108
31-Research and development	0.1	962	2	97,554
32-Other business activities	7.7	117,322	22	1,073,095
35-Health and social work	0.1	1,650	1	48,777
Total cost	100	1,520,736	100	4,877,704
With 12% cost reduction	100	1,338,248	–	–

Table 6.3 Disaggregation by economic sectors (in %) of total Chilean employment in 2003

Economic Sector	Percentage of Total Employment
Mining and quarrying	1.4
Agriculture, fishing, and hunting	13.7
Industry	14.2
Electricity, water, and gas	0.4
Construction	8.0
Trade	19.5
Transport and communications	8.4
Financial services	8.2
Community and social services	26.3
Total	100.0

6.3.2 Environmental Vectors Data Source

With regards to energy consumption, most data come from the Chilean National Emissions Inventory (NEI) (Poch ambiental and Deuman, 2008). For the disaggregation of all the 37 the economic sectors of the IOT, the National Energy Balance (BNE) from the National Energy Commission (CNE) of Chile

Table 6.4 Proportion of total Chilean energy consumption and CO_2 emissions among economic sectors in 2003

Economic Sector	% of Total Energy Consumption	% of Total Co_2 Emissions
Energy industry	31.3	30.7
Manufacture industry, construction and mining	22.1	23.6
Transport	37.6	37.7
Fishing	0.8	0.8
Public, residential and commercial	8.2	7.2
Total	100.0	100.0

in 2003 has been consulted (Ministerio de Energía, 2003), except for the fishing and transport sectors, which the NEI has exclusive data for them.

Finally, due to the large contribution of fossil fuels in the Chilean energy mix, the distribution of the CO_2 emission vector among IOT economic sectors is based on the energy consumption vector.

Table 6.4 shows the distribution of the energy consumption and CO_2 emissions among economic sectors from official Chilean NEI.

6.4 Results and Discussion

The results in terms of the impacts throughout the whole life time of the plant are presented below distinguishing between socioeconomic, environmental, and social direct effects (DE) and indirect effects (IE) in both investment and O&M phases.

6.4.1 Socioeconomic Impacts: Direct, Indirect, and Induced Economic Effects and Employment Effects

Figure 6.1 shows the total direct, indirect, and induced economic effects (in US$M) of each economic sector[2] generated by the project.

The total effect that the solar plant would generate during its whole lifetime in the Chilean economy is 3,124 US$M; being 1,407 US$M direct effect; 1,683 US$M indirect effect and 34 US$M induced effects. The multiplier effect would be 2.19, indicating that for each dollar invested in the project, it would generate 2.19$ in the whole economy[3].

[2]The complete economic sector titles can be consulted in Appendix A (United Nations database).

[3]Comparing these results with the economic effects of the 17 MW Spanish plant, the total effect is 2,230 M€, with a multiplier effect of 2.3 (Caldés et al., 2009).

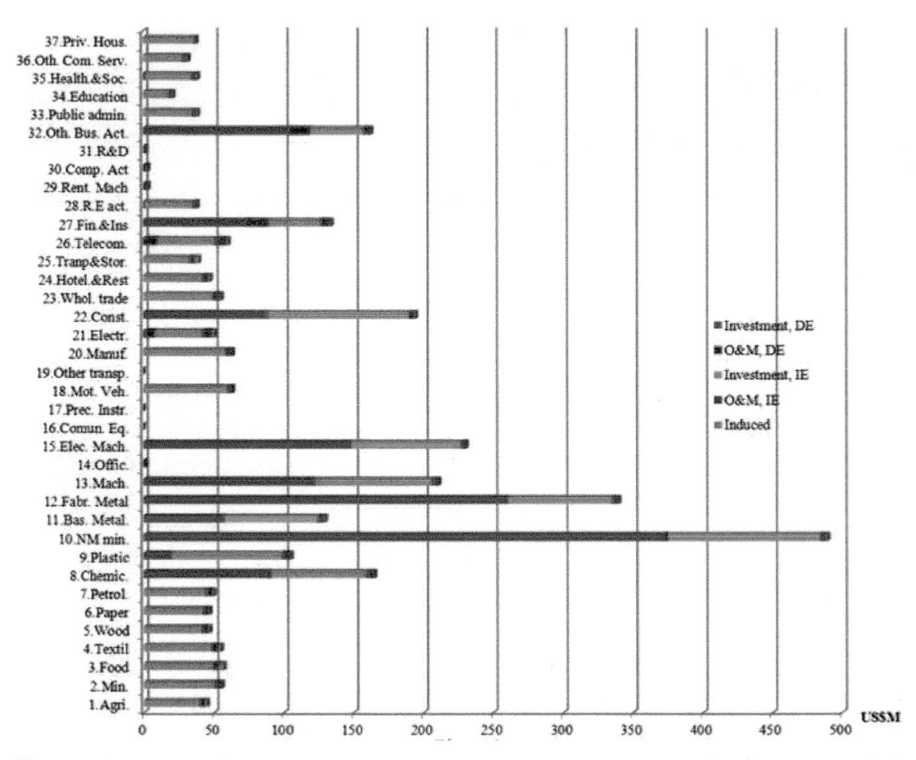

Figure 6.1 Direct (DE), indirect (IE), and induced economic effects (in US$M) in the whole life time of the plant in both investment and O&M phases by economic sectors.

Regarding direct effects, the main economic sector that would be affected in the investment phase is sector 10 “Other non-metallic minerals”. This fact is mainly due to the heliostats and the tower, which are the most expensive components of the plant. The second most benefited sector is sector 12 “Fabricated metals”, followed by the sector 8 “Chemicals”, sector 13 “Machinery” and sector 15 “Electrical machinery”. Other stimulated sectors would be sector 22 “Construction”, sector 27 “Financial intermediation”, and sector 32 “Other business activities”.

In the O&M phase, the largest stimulated sectors would be sector 27 “Financial intermediation” and sector 32 “Other business activities”, followed by sector 21 “Electricity, gas and water supply” and sector 26 “Post and telecommunications”.

Regarding indirect effects in both phases, the most affected sectors would be sector 10 “Non-metallic minerals”, sector 12 “Fabricated metals”, sector 11

"Basic metals", sector 13 "Machinery", sector 15 "Electrical machinery" and sector 22 "Construction[4]".

The induced effect[5] would be 33.4 US$M, being 15.4 direct effect and 18 indirect effect. In this case, the most affected sectors would be sector 3 "Food products, beverages, and tobacco" and sector 25 "Transport and storage" because of the large Chilean household expenditure in these sectors.

Referring to employment, the plant would create 134,949 new jobs in the whole life time of the plant, being 61,063 directs jobs and 73,887 indirect jobs. The ratio between indirect jobs over direct jobs would be 1.21 which, taking into consideration the labor force of the country, is consistent with similar results from the literature (Dii, 2011). Figure 6.2 shows the total direct and indirect full-time employments of 1-year of duration (in thousands employees) generated over the whole lifetime of the project in both investment and O&M phases by economic sector.

The sector[6] with the largest direct job creation in the investment phase would be sector 15 "Electrical machinery and apparatus" due to the large contribution of this sector in the power block (80%). Other sectors that would be affected but in less proportion are sector 10 "Non-metallic minerals"; sector 11 "Basic metals"; sector 12 "Fabricated metals"; sector 13 "Machinery and equipments"; sector 22 "Construction" and sector 32 "Other business activities" (mainly engineering jobs).

Regarding direct jobs in the O&M phase, the most stimulated sectors would be sector 32 "Other business activities" mainly related to engineering jobs; and sector 26 "Post and telecommunications".

The highest indirect employment would be also in sector 15 "Electrical machinery and apparatus". To a small extent, other sectors like sector 1 "Agriculture" or sector 4 "Textile products" would also be stimulated due to the relevance they have in the labor force of the country[7].

It is important to note that if imports had been taken into account; the domestic stimulation, both in terms of economic generation and employment

---

[4]The sectors that do not result indirectly stimulated are those which do not have data in the original IOT.

[5]The induced effect is only estimated in the O&M phase due to personnel cost data lack associated to the investment phase.

[6]Sectors that do not appear in figures is due to the insignificance of results.

[7]Some desegregation like "industry" have been done with the employment vector of Brazil because of Chilean data lack and regarding that the labor productivity in Brazil in 2003 was smaller (World Bank database), the employment results in some industrial sectors in Chile could be overestimated.

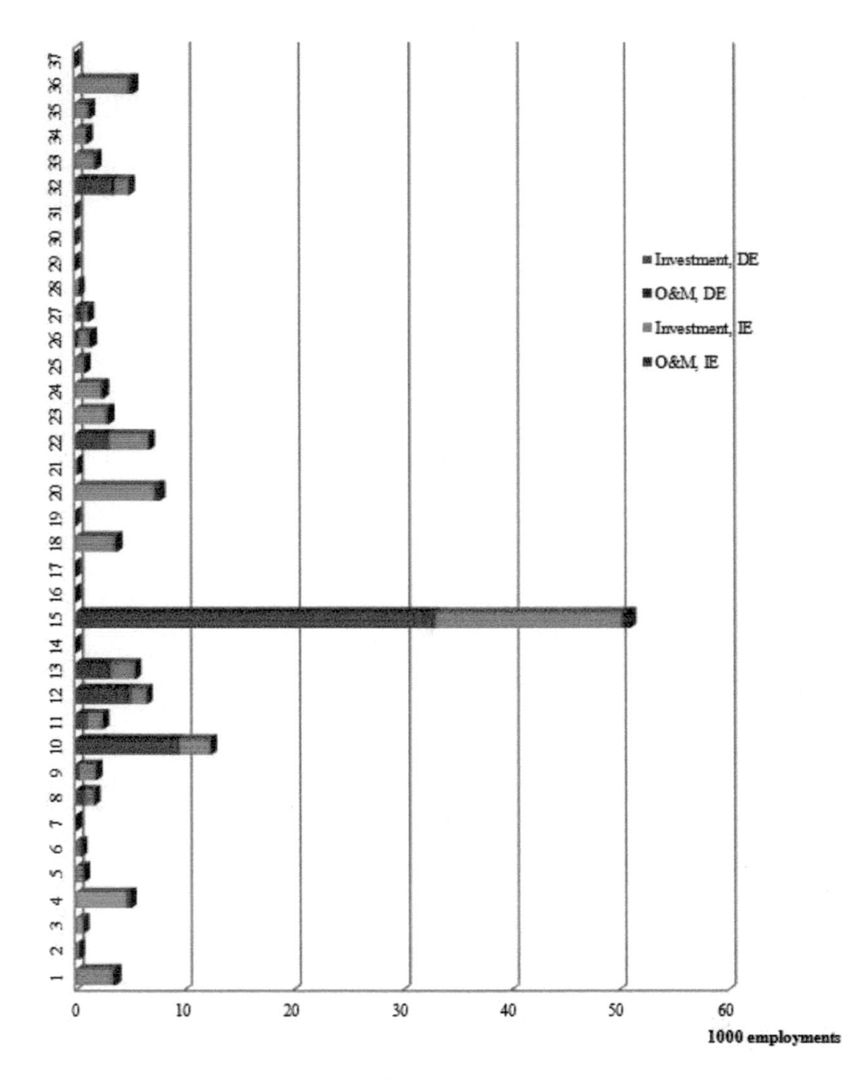

Figure 6.2 Direct (DE) and indirect (IE) employments in (thousand employments) in the whole life time of the plant in both investment and O&M phases by economic sectors.

creation, would have been smaller because some of the components (such as the receptor or the generator) are currently made in some specific countries and their manufacturing in a domestic scale is not expected at least in the short term (Dii, 2011). Nevertheless, it is expected that even when considering such imports, a remarkable stimuli to the domestic economy and job creation would take place (both directly and indirectly). This would lead to diversify business activities and contribute to the decoupling of the Chilean GDP from some key sectors such as mining, especially remarkable in the Northern part of the country.

6.4.2 Environmental Impacts: Energy Consumption and CO_2 Emissions

Figure 6.3 shows the direct and indirect energy consumption in thousands TJ of the whole life plant time in both investment and O&M phases by economic sectors.

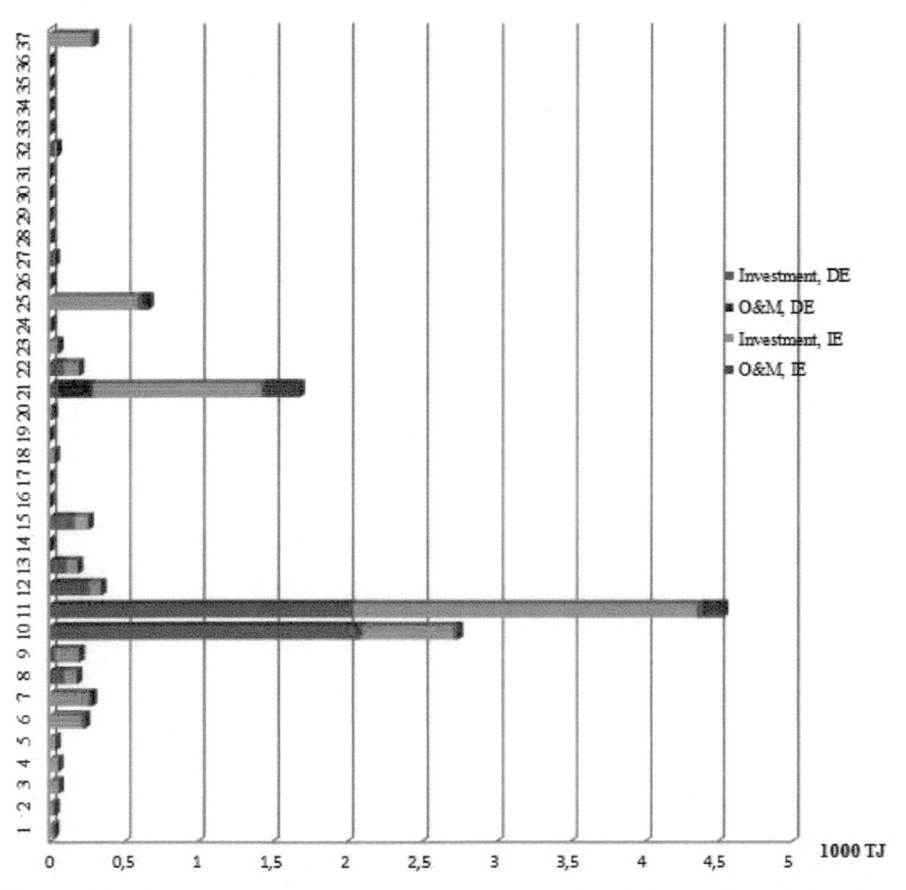

Figure 6.3 Direct (DE) and indirect (IE) energy consumption (in thousands TJ) in the whole life plant time in both investment and O&M phases by economic sectors.

The energy consumption throughout the whole life time of the project would be 12,243 TJ, being 5,225 TJ direct consumption, and 6,989 TJ indirect consumption.

In the investment phase, the largest consumption would come from sectors 10 "Non-metallic minerals" and 11 "Basic metals" due to the huge amount of energy they need for their production. In the O&M phase, the most affected sector would be sector 21 "Electricity, gas and water supply".

The indirect energy consumption in the investment phase would mostly come from sectors 11 and 21 "Basic metals" and "Electricity, gas and water supply"; followed by sector 6 "Paper products", sector 8 "Coke, refined petroleum products and nuclear fuel", sector 25 "Transport and storage" and sector 37 "Private households with employed people". In the O&M phase, the most affected sectors would be the most energy demanding sectors: 11, 21, and 25 'Basic metals', 'Electricity, gas and water supply' and 'Transport and storage'.

Regarding CO_2 emissions, the project would generate in the whole life time 1,759 Gg CO_2, being 1,007 direct emissions and 753 indirect emissions. The amount of emissions per kWh[8] would be 64.36 g CO_2 per kWh[9]. Figure 6.4 shows the direct and indirect CO_2 emissions in hundreds Gg of the whole life plant time in both investment and O&M phases by economic sectors.

Sectors with highest direct emissions in the investment phase would be sectors 10 and 11 "Non-metallic minerals" and "Basic metals" due to their large energy consumption based from the fossil fuel intensive Chilean energy mix. On a lesser extent, sectors 12 "Fabricated metals" and 15 "Electrical machinery" also would produce considerable amounts of emissions. In the O&M phase, the emissions would mainly come from sector 21 "Electricity, gas and water supply".

The main sectors that contribute to the indirect emissions coincided with the ones of the direct emissions, plus sector 25 "Transport and storage" and 37 "Private households".

Based on these results, it would be advisable to pay attention to these sectors and try to implement specific measures (e.g., renovation of obsolete components, technologies with CO_2 sequestration ...) and design policies aimed at improving their energy efficiency and reduce CO_2 emissions.

---

[8]It is assumed that the plant would produce 27,300 GWh in the whole life time.

[9]Some comparisons for this result have been done: 60.1 g CO_2/kWh from a 50-MW parabolic trough plant without natural gas consumption (De la Rua Lope, 2009); 60 g CO_2 eq/kWh (Lenzen, 1999) and 48 g CO_2 eq/kWh (Norton et al.,1998) from a solar tower plant.

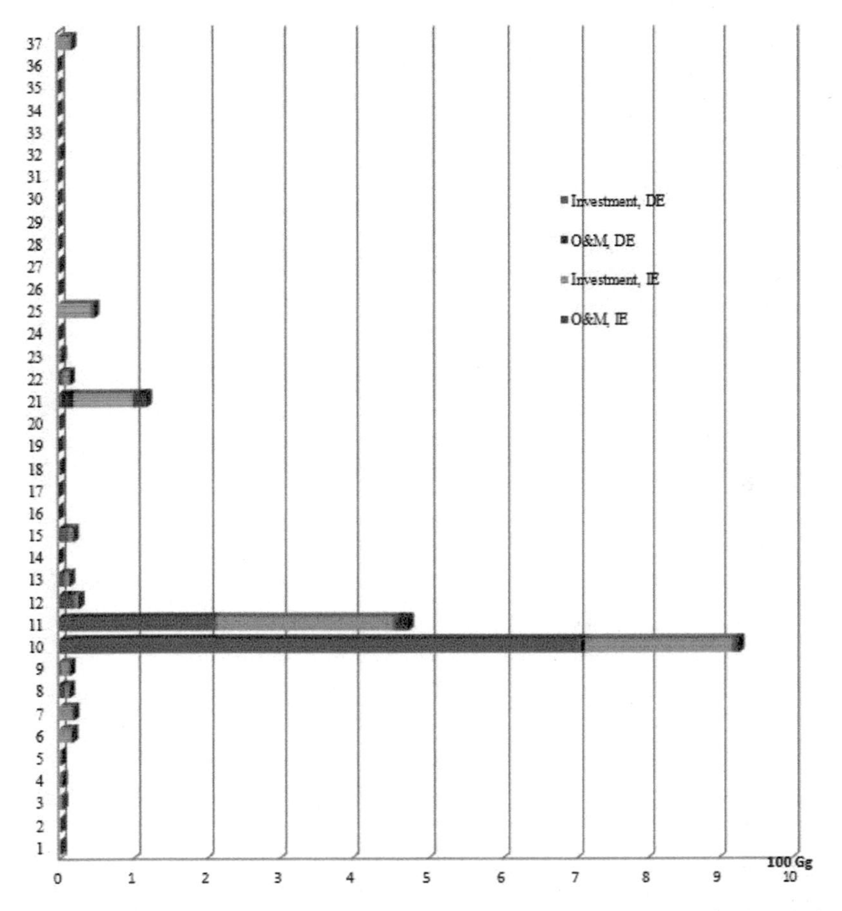

Figure 6.4 Direct (DE) and indirect (IE) CO_2 emissions (in hundred Gg) in the whole life plant time in both investment and O&M phases by economic sectors.

6.4.3 Social Impacts

The social risk analysis is performed in those sectors that show the largest total job creation as a result of the project. These sectors are sector 22 "Construction", sector 15 "Machinery and electric equipments", sector 20 "Manufacturing", sector 12 "Fabricated metals" and sector 10 "Non-metallic mineral products[10]".

[10]The correspondent sectors in the SHDB are "Construction", "Machinery and equipments nec", "Manufactures nec", "Metal products" and "Mineral products nec".

The SHI for these sectors has a value of 60.81, except in the construction sector which shows a higher value ($SHI_{const} = 67.5$). In Figure 6.5 it is possible to compare this sector value within different countries, in which the SHI in the construction sector in Chile is lower compared to some other Latin-American countries and even Spain, but higher compared to Norway. Also results differ among the different impact categories.

The higher SHI value of the Chilean construction sector compared to the rest of the Chilean sectors mentioned above comes from the Labor Rights and Human Rights categories. Within the Labor Rights category, the breakdown in each social theme,[11] is 5% forced labor, 19% freedom of association, 13% labor laws, 19% migrants workers, 5% poverty and 39% unemployment.

In this way, the most important risk in the construction sector in Chile is unemployment, with a risk value classification of "very high risk". This is consistent with some results from the Employment and Unemployment Survey in Santiago de Chile, which indicated that the construction sector had the largest unemployment rate (10.4%) in March 2014, doubling the country average unemployment rate (5.7%) (Centro de Microdatos del Departamento de Economía de la Universidad de Chile, 2014).

Additionally, the unemployment risk is also large in the rest of sectors, with a "high" risk value according to the SHDB.

In this sense, job creation as a result of the Cerro Dominador project could contribute to the decrease unemployment risks, especially in the construction sector as its "very high" risk value. This sector could be particularly relevant for the Chilean economy as it is one of the sectors that produce most of the economic and labor reactivation in the short and medium term. This is because its intensive labor force, mainly local, which, at the same time, stimulates the demand in various other sectors, both directly and indirectly (Solimano and Meller, 1983).

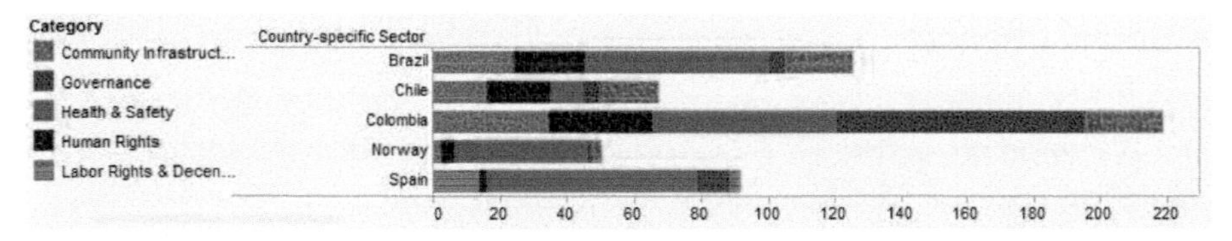

Figure 6.5 Comparison of the SHI in the construction sector and its different impact categories among different countries (Brazil, Chile, Colombia, Norway, and Spain).

[11]There are more social themes within the impact category "Labor rights and decent work" but they are not explained due to a data lack in the SHDB.

Local job creation as well as other livelihood effects as a result of CSP deployment in other areas of the word has been demonstrated and assessed (CSPToday, 2015). In particular, the results from this research in Ouarzazate area (Morocco) indicate that more than 1,500 new jobs were created in Morocco, of which 700 were generated locally in the Ouarzazate area. Additionally to these jobs, other social effects were assessed such as the reinforcement of families bonds related to migratory flows and new incomes, as well as creation of new infrastructures. However, this research also indicates that official entities like ministries should support local industries, promote skills development and research and development (R&D) in order to increase the competitiveness of the sector and increase productivity among the whole supply chain necessary for the project. Finally, despite all the advantages, the principal concerns of the studied population in Ouarzazate were related to water needs for the operation and maintenance of the plant in such an arid region, lack of local population capacity, immature industrial development and issues related to non-transparency and absence of local population participation in the decisions making (CSPToday, 2015). Such findings could be of great relevance for Chile as the local population might encounter similar opportunities and challenges.

6.5 Conclusion

Based on CSP cost data and the Chilean IOT, this paper has conducted a sustainability impact assessment of a CSP project in Chile in terms of economic stimulation, job creation, energy consumption, CO_2 emissions, and existence of social risks in the country.

Under a 100% local content assumption, the economic sectors most benefited by the project would be "Non-metallic mineral" and "Fabricated metals" sectors. Additionally to these sectors, this project would stimulate (both directly and indirectly) other domestic sectors like "Construction".

In terms of employment creation, the most stimulated sector is the one related to machinery and electrical equipment. Other sectors with large job creation figures would be "Non-metallic minerals", "Fabricated metals", and "Construction".

With regards to environmental impacts, those sectors with the largest energy consumption and CO_2 emissions would be "Non-metallic minerals" and "Basic metals".

Finally, the social risk assessment indicates that high unemployment is the most relevant social risk is those sectors most affected socioeconomically by the project. Therefore, the project would lead to social benefits. Results also indicate that other social risks (e.g., child labor) would also have to be monitored.

To conclude, this paper has shown that CSP technology could play a relevant role as a driver to stimulate and diversify the Chilean economy as well as to generate new employment opportunities, with a remarkable potential to reactivate the local economy. Additionally, the large solar potential together with the environmental benefits as a renewable technology, indicate that CSP could be a promising technology in Chile in a short-medium term, specifically in a region with high solar irradiation like Atacama Desert. Consequently, support policies that account for such positive externalities should be put in place in order to foster the deployment of this technology in Chile.

Acknowledgments

This work has been funded by the grant of CIEMAT for the training of research personnel.

Appendix A: Description of Chilean IOT (United Nations)

Sector 1: Agriculture, hunting, forestry and fishing
Sector 2: Mining and quarrying
Sector 3: Food products, beverages and tobacco
Sector 4: Textiles, textile products, leather and footwear
Sector 5: Wood and products of wood and cork
Sector 6: Pulp, paper, paper products; printing and publishing
Sector 7: Coke, refined petroleum products and nu Sector 1: clear fuel
Sector 8: Chemicals and chemical products
Sector 9: Rubber and plastics products
Sector 10: Other non-metallic mineral products
Sector 11: Basic metals
Sector 12: Fabricated metal products except machinery and equipment
Sector 13: Machinery and equipment n.e.c
Sector 14: Office, accounting and computing machinery
Sector 15: Electrical machinery and apparatus n.e.c
Sector 16: Radio, television and communication equipment
Sector 17: Medical, precision and optical instruments
Sector 18: Motor vehicles, trailers and semi-trailers
Sector 19: Other transport equipment
Sector 20: Manufacturing n.e.c; recycling
Sector 21: Electricity, gas and water supply

Sector 22: Construction
Sector 23: Wholesale and retail trade; repairs
Sector 24: Hotels and restaurants
Sector 25: Transport and storage
Sector 26: Post and telecommunications
Sector 27: Finance and insurance
Sector 28: Real estate activities
Sector 29: Renting of machinery and equipment
Sector 30: Computer and related activities
Sector 31: Research and development
Sector 32: Other Business Activities
Sector 33: Public admin. and defence; compulsory social security
Sector 34: Education
Sector 35: Health and social work
Sector 36: Other community, social and personal services
Sector 37: Private households with employed persons

References

[1] ABENGOA. (2013). *Declaración de Impacto Ambiental de Cerro Dominador*. Available at: http://es.slideshare.net/VctorA1/adenda-n-1planta solarcerrodominador [cited 2015 Jun 30].

[2] Avila-Marin, A. L., Fernandez-Reche, J,, Tellez, F.M. (2013). Evaluation of the potential of central receiver solar power plants: configuration, optimization and trends. *Appl. Energy* 112, 274–288. doi: 10.1016/j.apenergy.2013.05.049

[3] Banco Central de Chile. (2012). *Base de Datos Estadísticos – Precio Interior Bruto Regional por Sectores Económicos en 2012*. Available at: http://si3.bcentral.cl/Siete/secure/cuadros/arboles.aspx [cited June 30, 2015].

[4] Caldés Gómez, N., and Lechón Pérez, Y. (2010). *Análisis de externalidades de las energías renovables [Internet]. Tratado de energías renovables*. Editorial Aranzadi. p. 951–1004. Available from: http://dialnet.unirioja.es/servlet/articulo?codigo=3187593 [cited June 30, 2015].

[5] Caldés, N., Varela, M., Santamaría, M., and Sáez, R. (2009). Economic impact of solar thermal electricity deployment in Spain. *Energy Policy* 37, 1628–1636.

[6] Centro de Microdatos del Departamento de Economía de la Universidad de Chile. (2014). *Encuesta de Ocupación y Desocupación en el*

Gran Santiago: Infrome Trimestral de Empleo Marzo. Deparmento de Economía, Santigo, CL.

[7] Chile Central Bank. (2009). *Clasificación del gasto en consumo final de los hogares e instituciones privadas sin fines de lucro por finalidad, periodo 2003–2007*. Banco Central de Chile, Santigo, CL.

[8] Comisión Nacional de Energía. (2012). *Electricidad – Generación Bruta SING en 2013*. Available at: http://www.cne.cl/estadisticas/energia/electricidad [cited June 30, 2015].

[9] CSP World. *CSP World Map | CSP World* [Internet]. Available at: http://www.cspworld.org/cspworldmap [cited June 30, 2015].

[10] CSPToday. (2015). *Five things you need to know about community buy-in in Morocco*. Available at: http://social.csptoday.com/markets/five-things-you-need-know-about-community-buy-morocco [cited June 30, 2015].

[11] CSPToday. (2012). *Guía de Internacionalización de la CSP*.

[12] Dii. (2011). *The Economic Impacts of Desert Power: Socio-economic aspects of an EUMENA renewable energy transition*. Dii, München.

[13] Fichtner. (2010). *Assessment of Technology Options for Development of Concentrating Solar Power in South Africa*. The world Bank, Johennesberg.

[14] GreenDelta. *Social hotspot database Introductory User Tutorial*.

[15] De Gregorio, J. (1998). *Comportamiento de los agentes económicos*.

[16] Hernández-Moro, J., Martínez-Duart, J. M. (2013). Analytical model for solar PV and CSP electricity costs: present LCOE values and their future evolution. *Renew. Sustain. Energy Rev.* 20, 119–132.

[17] Hoffschmidt, B., Alexopoulos, S., Rau, C., Sattler, J., Anthrakidis, A., Boura, C., et al. Concentrating Solar Power. *Compr. Renew. Energy* 3, 595–636. doi: 10.1016/B978-0-08-087872-0.00319-X

[18] Holland, D, and Cooke, S. C. (1992). Sources of structural change in the Washington economy. *Ann. Reg. Sci.* 26, 155–170. doi: 10.1007/BF02116367

[19] Instituto Nacional de Estadísitica. (2003). *Series Empalmadas Diciembre-Febrero 1986 a Diciembre-Febrero 2010 – Base Censo 2002 | Instituto Nacional de Estadísticas | INE 2014*. Available at: http://www.in e.cl/canales/chile_estadistico/mercado_del_trabajo/empleo/seri es_estadisticas/nuevas_empalmadas/series_fecha.php [cited Jun 30, 2015].

[20] Kammen, D. M., and Dove, M. R. (1997). The Virtues of Mundane science. *Environ. Sci. Policy Sustain. Develop*. 39, 10–41.

[21] Kolb, G., Ho, C., Mancini, T., and Gary J. (2011) *Power tower technology roadmap and cost reduction plan. SAND2011-2419, Sandia.* Available from: http://prod.sandia.gov/techlib/access-control.cgi/2011/112419.pdf

[22] De la Rua Lope, C. (2009). Desarrollo de la herramienta integrada "análisis de ciclo de vida – Input Outout análisis para España y aplicación a tecnologías energéticas avanzadas." Planta.

[23] Larraín, T., and Escobar, R. (2012). Net energy analysis for concentrated solar power plants in northern Chile. *Renewable Energy* 41, 123–33. doi: 10.1016/j.renene.2011.10.015

[24] Lenzen M. (1999). Greenhouse gas analysis of solar-thermal electricity generation. *Solar Energy* 65, 353–68.

[25] Leontief, W. (1936). *Quantitative input and output relations in the economic systems of the United States.* Available from: http://www.jstor.org/stable/1927837?seq=1#page_scan_tab_contents [cited June 30, 2015].

[26] Linares, P., Leal, J., Sáez R. (1996). *Evaluación de las externalidades de la biomasa para producción eléctrica.* Available at: http://bddoc.csic.es:8080/detalles.html?id=113934&bd=ICYT&tabla=docu [cited June 30, 2015].

[27] Ministerio de Energía. (2003) *Balance Nacional de Energía 2003.* Available at: http://antiguo.minenergia.cl/minwww/opencms/14_portal_informacion/06_Estadisticas/Balances_Energ.html [cited June 30, 2015a]

[28] Ministerio de Energía. (2013). *Gobierno promulga Ley 20/25 y anuncia entrada en vigencia de Ley de Concesiones.* Available at: http://www.minenergia.cl/ministerio/noticias/generales/gobierno-promulga-ley-20-25-y-anuncia.html [cited June 30, 2015c]

[29] Ministerio de Energía. Energía 2050 – Proceso Participativo Política Energética – Ministerio de Energía – Gobierno de Chile. 2014 [Internet]. [cited 2015b Jun 30]. Available from: http://www.energia2050.cl/programa

[30] Ministerio de Energía. Ley 20.257: Modificaciones a la ley general de servicios eléctricos respecto de la generación de energía eléctrica con fuentes de energías renovables no convencionales. 2008;2008(7681): 7681.

[31] Norton, B., and Eames, P. C., and Lo, S. N. (1998). Full-energy-chain analysis of greenhouse gas emissions for solar thermal electric power generation systems. *Renew. Energy* 15, 131–136.

[32] Organization for Economic Cooperation and Development. (2003). *Trade, Input–Output Tables.* Available at: http://www.oecd.org/trade/input-outputtables.htm. [cited June 30, 2015].
[33] Poch ambiental and Deuman. (2008). Inventario nacional de emisiones de gases de efecto invernadero.
[34] Ten Raa T. (2006). *The Economics of Input-Output Analysis*. Cambridge University Press, Cambridge Available at: /ebook.jsf?bid=CBO9780511610783. [cited June 30, 2015].
[35] Seminario Iberoamericano de Energías Renovables. (2009). *Las energías renovables en América Latina: Chile*. Available at: http://es.slideshare.net/CanalEndesa/las-energas-renovables-en-amrica-latina-chile. [cited June 30, 2015].
[36] Solimano, A., and Meller, P. (1983). *Desempleo en Chile: interpretación y políticas económicas alternativas.*
[37] Social Hotspot Database. Available at: http://socialhotspot.org/ [cited June 30, 2015].
[38] Tarancón MÁ. (2003). Técnicas de Análisis Económico Input–Output.
[39] United Nations database. *United Nations Statistics Division – Classifications Registry.* Available at: http://unstats.un.org/unsd/cr/registry/regcs.asp?Cl=17&Lg=1&Co=23 [cited June 30, 2015].
[40] World Bank database. *Indicators | Data.* Available at: http://data.worldbank.org/indicator [cited June 30, 2015].
[41] World Commission on Environment and Development. (1987). Our Common Future (The Brundtland Report). *Med. Confl. Surviv.* 4, 300.
[42] World Input Output Database. *WIOD Data.* Available at: http://www.wiod.org/new_site/database/seas.htm [cited June 30, 2015].

7

World Renewable Energy Congress and Network: WREC–WREN Activities 2014 and 2015

Ali Sayigh

WREN, PO Box 362, Brighton BN2 1YH, UK
E-mail: asayigh@wrenuk.co.uk

Mission Statement

With the accelerated approach of the global climate-change point-of-no-return the need to address the pivotal role of renewable energy in the formation of coping strategies, rather than prevention, is more crucial than ever. Sustainability, green buildings, and the development of the large-scale renewable energy industries must be at the top of all development, economic, financial, and political agendas. The time for action has arrived. Prevention and questioning how and why we face this great challenge is a luxury we can no longer indulge. The recent Conference **13 December 2015 at Paris Agreement COP21** to limit CO_2 so that the temperature will not rise more than 1.5°C is a world significant achievement from 190 countries, Figure 1. WREC and WREN since their establishment in 1990 have strived to promote renewable energy through conferences, seminars, symposia, books and journals throughout the world.

Renewables are the cornerstone and the foundation of a truly sustainable energy future. Our mission is to promote enabling policies and to further develop a broad range of renewable energy technologies and applications in all sectors – for electricity production, heating and cooling, agricultural applications, water desalination, industrial applications, and for the transport sectors leading to A BETTER, CLEANER AND SAFER WORLD

Figure 1 The largest Global Revolution on 13 December 2015 at Paris Agreement COP21, stating.

1. The earth temperature should not increase by more than 1.5 °C,
2. Each country pledged to make zero emission in their century,
3. Creation of annual fund US$1000 billion to help those effected by climate change,
4. Check on progress in 5-year intervals, signed by 180 countries including USA, China, and Japan.

World Renewable Energy Congress WREC XIII: Renewable Energy in the Service of Mankind (03–08 August 2014, University of Kingston, London, UK)

Delegates, Experts, Invited Speakers, and Contributors started to arrive at Kingston-on-Thames from Friday 1 August. Among the delegates and VIPs were: H E Dr. Abdulaziz O. Altwaijri – Director General of ISESCO which represents 57 countries; Mr. Hans-Josef Fell – DWR eco GmbH, Berlin, Germany; Prof. Dr. Mohamed El-Ashry, Senior Fellow, UN Foundation, USA; Prof. Brian Norton, President of Dublin Institute of Technology and Ex-President of WREN; Prof. Dr. Ebrahim Al-Janahi, President of Bahrain University; Prof. Dr. Soogab Lee, President of Korean Wind Energy Association and Deputy Vice Chancellor of Seoul National University; Dr. Lawrence Kazmerski,

Ex-Director of Photovoltaic, NREL with Dr. Charles F. Kutscher, Director of Buildings and Thermal Systems Center, NREL, USA, and Dr. David Renne, President of ISES.

On 3 August, prior to WREC XIII opening, the WREN Council meeting took place with 70 participants from 47 countries. Prof. Ali Sayigh detailed the WREN activities of the last two years and reflected on the coming 2 years when cooperation and promotion of renewable energy will take place in particular with Malaysia, Romania, Italy, and the World Renewable Energy Congress – WREC XIV which will be held in September 2016 in Indonesia. Dr. Herliyani Suharta gave a talk on the arrangements for WREC XIV Indonesia.

This was followed by a presentation from Dr. Ruxandra Crutescu (Figure 2) concerning World Renewable Energy Congress – Romania, June 8–13 2015.

Figure 2 Dr. Ruxandra Crutescu – Chair of Romanian Organizing Committee with Prof. Nader Al-Bastaki, Engineering College, Bahrain University.

This was followed by a welcome to Florence from Prof. Ali Sayigh and Prof. Marco Sala, and also the announcement of the third Mediterranean Green Buildings and Renewable Energy Forum – MGBARE, 23–26 August 2015 at University of Florence, Florence, Italy. Figure 3: Prof Marco Sala Director of ABITA, Florence, Italy.

Figure 4 Dr Suharta is the Chair of the Technical Committee while Prof Herman Ibrahim is the Chairman of the Organizing Committee for WREC – XV, September 2016.

WREC XIII details:

There were 550 participants from 95 countries. The number of abstracts was close to 600. The Congress Opening Ceremony took place on 4 August. A Plenary program took place each morning and was divided into two parts: Main Plenary from 9:00–10:30 and two parallel sessions from 11:00–13:00. Every afternoon from 13:00–18:30 there were 11 parallel Technical Sessions.

The opening session took place at the Rose Theatre in the city and after a brief welcome by Prof. Sayigh, the Vice-Chancellor of Kingston University, Prof. Julius Weinberg, welcomed the participants to the University and wished them a very successful Congress. Then Prof. Edith Sim, Honorary Chair of the Congress, added her welcome to the participants. Prof. Norton outlined the mission of WREN and stressed the importance of the organization in promoting renewable energy globally and creating a forum for networking among countries and individuals. He described briefly the history of WREN. This was followed by H E Dr. Abdulaziz Bin Othman Altwaijri, Director General of ISESCO who gave a speech outlining the mission and goals of ISESCO in promoting energy and renewable energy among its 57 member states and the rest of the world. H E explained how ISESCO has cooperated with WREN over the last 15 years.

Prof. Barbara Pierscionek, Associate Dean for Research, University of Kingston spoke of the creation of Kingston Institute of Energy and its role in promoting research in all aspects of energy.

Then The Rt. Hon. Edward Davey MP, Secretary of State for Energy and Climate Change delivered a speech outlining the UK Program in Energy and Climate change which was followed by an open discussion,

H E Dr. Abdulaziz Altwaijri at the Opening

VC Prof Julius Weinberg

The Rt. Hon. Edward Davey MP, Secretary of State for Energy and Climate Change.

WREC – XIII in London by Prof. David Elliott

Some critics continue to portray renewables as marginal, with for example, ExxonMobil claiming that their potential is limited by '*scalability, geographic dispersion, intermittency (in the case of solar and wind), and cost relative to other sources*', with renewables only likely to make up about 5% of the global energy mix by 2040.

www.ft.com/cms/s/0/5a2356a4-f58e-11e3-afd3-00144feabdc0.html?siteedition=uk#axzz33albsQ2B

Most, however, see renewables as booming, with IRENA looking to 30% or more of primary energy globally by 2030 (www.irena.org/remap). That is the sort of future envisaged, on the way to maybe near 100% of power by 2050, by most who attended the 13th biannual **World Renewable Energy Congress**, this one at Kingston University, London, in August. As usual, it kicked off with the big hitters laying out overviews, in the plenary sessions, with the '100% renewables' theme being to the fore. For example, Stefan Schurig, from the **World Future Council Foundation**, described their 100% renewables campaign (www.go100re.net) and David Renné, President of the **International Solar Energy** Society looked at 'A 100% Renewable Energy Future', which he saw as 'a solution to climate change'. He stressed the need for grid balancing. More generally Hans-Josef Fell, President of the German **Energy Watch Group**, looked at the role of renewables in 'global cooling', and Mohamed El-Ashry from the **United Nations Foundation**, also spoke very positively on the future of renewables – and REN21's work: http://www.ren21.net/REN21Activities/GlobalStatusReport.aspx

Key regional and national prospects were also covered. Rainer Hinrichs-Rahlwes, Vice-President of the **European Renewable Energies Federation** looked at 'Perspectives for Renewable Energy in Europe', and then, in his **German Renewable Energy Federation** (BEE) role, at 'Perspectives of

Renewables in Germany'. He was critical of the **German** governments recent plans to dump the Feed In Tariffs and cut support levels and targets, asking whether the new program was 'reform or deform'. Prof. Abubakar Sambo looked at developments in Africa: he talked of getting to 50% renewable energy after 2040. Prof. Yogender Kumar Yadav looked at options in **India**, including 100 GW each of wind and PV, plus a lot of bioenergy: he looked to bio-refineries becoming a major industrial focus.

It was an impressive set of overviews, reflecting the ability of WREC, and its **WREN** network, to bring together people from across the world, a strength celebrated in the introductory talk by long time WREN member, Prof. Brian Norton, President, Dublin Institute of Technology, who reflected on its first 25 years. But its strength is not just in big names and high policy. In parallel, WREC had a vast number of technical papers and regional reports covering just about all aspects of renewables across the world, with many strong presentations on North Africa (a new growth area), the Middle East, and Asia – China, India, and Indonesia especially. So it's timely that the next WREC (No 15) will be in Indonesia in 2016.

The coverage was certainly very international, with for example 20 papers from Algeria, 10 from Egypt, 6 each from China, India and Indonesia, 16 from Japan, 17 from S Korea, 10 each from the USA, Canada, and Australia, plus 10 from Africa and 20 from South America. However, as the host country, there were inevitably many UK papers, 74 out of the near 500 total. There certainly is a lot going on here and in his introductory remarks, Energy Secretary (and local MP) Ed Davey naturally talked the UK efforts up and launched Kingston Energy, a new initiative by Kingston University: http://sec.kingston.ac.uk/energy/

My own contribution was more critical, contrasting the relatively slow growth and policy constrained future with the possibilities I and David Finney had identified in the Pugwash 80% Renewables UK 2050 Pathway. A similarly critical view was put forward by Martin Alder, Managing Director of renewable developer Optimum Energy, in an overview of 'UK Government Support for Renewables 2010 to the Present'. He concluded that the UK renewables program has become politicized, with 'the previous consensus evaporating' and the continued expansion of renewables being under threat. At the very least he predicted a 'hiatus' while the new CfD process, and the new funding cap on it, was digested- most developers would wait to see how that went and stick to the Renewables Obligation for now.

There was a large non-UK European contingent, with 23 papers from Italy, 18 from Germany, while Spain had 17, Greece 16, France 11, and Denmark 9.

Amongst the most intriguing of these was one by Avril et al., from CEA Saclay, France: 'Nuclear Power: a promising back-up option to promote renewable penetration in the French power system?' It argued that since France had such a large nuclear capacity (63 GW, likely to fall to maybe 50 GW soon given the new partial phase out policy), even if only a small percentage change in output could be made regularly and fast, that would be enough (maybe 5 GW) to balance a large renewable component. The implication was that since the capacity existed it might as well be used and the old, fully paid for, nuclear plants could act as grid balancing units at less cost than new gas fired turbines, depending on the price of carbon. Clever stuff. It is usually argued that nuclear plant can't load follow to any great extent quickly and repeatedly, since there is a problem with radioactive Xenon poisoning (it takes time to clear). But they can make smaller adjustments. Of course no one would suggests building new nuclear plants to balance renewables – you would have to have a lot to be able to do that. And at present it is not planned that any of the new plants proposed for the UK will load follow. But France is unique in that it has a lot, some of which can and do load follow. The plants are still be basically inflexible, but they could help a bit with some real time balancing, and, in addition, the paper argued, some of their excess output (e.g., at night) could be converted to hydrogen to use more flexibly to balance renewable short falls at other times. That idea has of course already been taken up for wind surpluses (i.e., wind to gas). Could it be that, where they still exist in significant numbers, nuclear plants will muscle in to the balancing market as well?

For most of those attending WREC, this sort of issue would no doubt be irrelevant; their focus is on getting renewables established rapidly, whether in Europe, led by Denmark, where it was reported that wind was now supplying 33.2% of electricity and was expected to reach 50% by 2020, and 70% by 2050, or (increasingly) in the developing world. As usual WREN produced a *Renewable Energy* review report for the conference. It included a global overview, claiming that it was possible to get at least 50% of global electricity from renewables by 2030, produced by Prof. Ali Sayigh, WREC and WREN's ever energetic chair. Gatherings like this may help to make that, and maybe more, a reality. www.wrenuk.co.uk

There were three major dinners during the Congress. First one was on Monday 4 August which was sponsored by SPRINGER Publishing Company. Prior to the Dinner WREN presented 2014

Pioneers' awards to the recipients from Malaysia, India, Korea, Australia, Sweden, Germany, and UK. The awards were produced in Malaysia and donated by University of Kebangsaan Malaysia and the medals were donated by University of Bahrain. Dr. Larry Kazmerski, Prof. Al-Bastaki, and Prof. Sayigh presented the awards.

Prof. Yogender Kumar Yadav – India

Prof. Soogab Lee and Mrs. Lee – Korea

Prof. Giuliano Premier UK

Also among those who received awards, were H E Dr. Altwaijri, ISESCO, Ms. Tiffany Gasbarrini, Springer, and Dr. Hossein Mirzaii, Executive Vice Chair of WREC XIII, Kingston University.

H E Dr. Altwaijri receiving the outstanding Medal for the promotion of Renewable Energy within 57 Countries of ISESCO Organization and Bahrain University Shield.

Dr. Hossein Mirzaii receiving Bahrain University Shield in appreciation of his enormous contribution to the Organization of the Congress in Kingston University.

Ms. Tiffany Gasbarrini – Springer receiving the outstanding medal for her continued support to renewable energy & WREN.

During the Dinner a talk was delivered by Mr. James Watson – Executive Director of EPIA, on the Role of EPIA in promoting the use of PV in its various industrial members.

This was followed by Special announcement by Springer in establishing a new Journal called: **Renewable Energy Sustainability and Technology – REST** with Prof. Ali Sayigh, Editor-in-chief.

The dinner was attended by 105 participants and was a very cordial evening. Our thanks go to Springer, Malaysia and Bahrain and most of all to the Master of Ceremonies Dr. Larry Kazmerski.

On 5 August, a dinner was sponsored by Elsevier. There were 95 participants and Prof. Sayigh thanked Elsevier for their continued support to WREN and explained that WREN Journal – Renewable Energy is in the good hand with Prof Soteris Kalogirou who had been recently appointed as Editor-in-chief. Then Ms. Lisa Reading welcomed all participants in the name of Elsevier. Then Prof. Kalogirou said he would do his best to keep up the standard of Renewable Energy and encouraged everyone to publish in the Journal. Prof. Robert Critoph then acknowledged Mrs. Linda Sayigh who during the last 14 years had been the main controller and executive editor of Renewable Energy.

The third dinner was the Gala Dinner on 6 August at Ravens Ait Island. This was attended by 165 guests who greatly enjoyed the delightful setting on this island in the river Thames and an excellent dinner.

During the dinner Prof. Wasim Saman from University of South Australia gave a talk about a solar village in Adelaide area.

At the closing ceremony on 8 August each head of the various Technical Committees gave a 5-min briefing about what happened in their section and where and what the advances and areas of research should be in the coming years.

Then Prof. Ali Sayigh/WREN Trophy, was presented to the acting Ambassador of Denmark and his assistant, Denmark was chosen by the technical committee as the country which had increased its renewable energy capacity to the greatest extent over the last two preceding years. Previous award winners were 2006-Cyprus, 2008-Brazil, 2010-Germany and 2012- Spain.

Mr. Christian Gronbech-Jensen Danish Minister Counsellor, Deputy Head of Mission

After this ceremony the remaining participants gathered in the courtyard of JG Building to take a collective photograph with the Danish acting Ambassador with WREN Trophy.

There was a book exhibition during the Congress. Five exhibitors were present and the first four of them were also sponsors: The Institute of Engineering and Technology Publishing (IET); SPRINGER publishing Company; Elsevier UK; and Gunt Technology Ltd.

Additional Photos are Below Related to Various Congress Activities and Events

Prof. Datto Sopian

Prof. Dr. Winfried Hoffmann

Prod Andrew Garrad

Prof. Soogab Lee

Dr. Michael Geyer

Dr. Stan Shire

Mr. Bill Watts

Prof. Saad Mekhilef, Prof. Ali Hamzeh, and Prof. Jerry D Murphy Organizing Committee

Other Activities in 2014
Other 2014 Events

The World Renewable Energy Congress and WREN participated at PVSEC – 29th September where Prof Sayigh chaired a session.

In Muscat, Oman, Prof. Sayigh gave a keynote presentation on renewable energy and security at the Oman National Conference On Energy Security 12–16 October.

Third from the left: Prof. Raof Bin Ahmed and Nejieb Mansouri – Pioneers in Renewable Energy.

In Tunis, at the Enersol World Sustainable Energy Forum on Renewable Energy, 27–28, November 2014 where Prof Sayigh gave a keynote address on ways in which Tunisia could cover all its electricity demand from photovoltaic devices.

World Renewable Energy Council/Network (WREN) Seminar No. 88 Renewable Energy: Policy, Security, Electricity, Sustainable Transport, Water Resources/Management and the Environment 07–13 December 2014, Old Ship Hotel, Brighton

The Seminar began on Sunday 7 December with the arrival of most of the delegates.

There were 24 participants and speakers from 14 countries.

Some of the participants

Participants at a session

On Wednesday, the participants had lunch at the residence of Prof. Donald and Mrs. Margo Swift-Hook in Woking, Surrey after which they went to London for their scientific visit to Kings Cross station with 250 kW photovoltaic panels. In the same evening the participants attended the David Hall Memorial lecture given by **Dr. Tim Bruton**, TMB Consulting, Milliwatts to Gigawatts – the PV journey.

Kings Cross Station

Memorial Lecture

In December, ISESCO held their Ministers of Higher Education Conference in Rabat, Morocco and Prof. Sayigh was one of the delegates.

H E Dr. Abdulaziz Altwaijri with H E Moroccan Minister of Higher Education honouring one of the participants from Malaysia.

Events in 2015

ISESCO – WREN – University of Bahrain
Seminar on:
Renewable Energy for Sustainable Development
15–18 February 2015, Crowne Plaza Hotel, Manama, Kingdom of Bahrain

Under the Patronage of H E the Minister of Energy of Bahrain Dr Abdul Hussain bin Ali Mirza on February 2015 at Hotel Crowne Plaza, H E the Minister of Energy opened the Seminar "the 7th International Seminar & Expert Meeting on RENEWABL ENERGY for SUSTAINABLE DEVELOPMENT" with a speech outlining Bahrain's commitment to sustainable development and the importance of renewable energy in the country and outlined the growth of electricity and water supply in Bahrain. H E thanked ISESCO and WREN for their continuing support of the Seminar. He wished the participants success in their deliberations. There were 100 or more participants among them 3 ministers, H E the President of Bahrain University Dr. Ebrahim Al-Janahi, Press and Television Media present at the Seminar.

H E Dr. Abdul Hussain bin Ali Mirza Bahrain Minister of Energy

H E the Minister before the opening of the Seminar

The Seminar was organized by the University of Bahrain and the World Renewable Energy Congress/Network (WREN) with major sponsorship from the Islamic Educational Scientific and Cultural Organization (ISESCO).

The Minister Dr. Mirza highlighted the need for and the advantages of using Renewable Energy in Bahrain. He expressed his thanks to ISESCO in sponsoring the Seminar and thanked and praised the cooperation between University of Bahrain, and WREN.

Both Dr. Fuad Al-Ansari – Dean of the College of Engineering, who gave the University of Bahrain welcome speech, and Dr. Ali Rahal thanked the organizers and sponsors of the Seminar. H E Dr. Abdulaziz bin Othman Altwaijry's sent greetings and good wishes for a successful meeting.

Finally, Prof. Ali Sayigh – chairman of WREC/WREN thanked H E Dr. Mirza and the University of Bahrain for organizing the Seminar for the benefit of the region. He also expressed his thanks and appreciation to H E Dr. Abdul Aziz bin Othman Al-Twaijri for the continuous support of WREN activities in the promotion of Renewable Energy and Clean Environment. Prof. Sayigh outlined the necessity for Bahrain to speed up its utilization of Renewable Energy for the betterment of the Bahrain and the GCC countries.

Dr. Fuad Al-Ansari, Dean of College of Engineering UOB

Dr. Ali Rahal – ISESCO Representative

Prof. Ali Sayigh Chairman of WREC/WREN

Group Photograph with H E The Minister of Energy Dr. Mirza

Aims of the Seminar

1. To make the audience aware of the importance and use of Renewable Energy (RE).
2. To identify the best forms and means of utilizing RE in the GCC Countries and in Bahrain in particular.
3. To explain the means of generating electricity from R E and putting the appropriate laws in place to introduce the feed-in-tariff (FIT) in Bahrain.
4. To learn about projects and program of R E in the GCC states and globally.
5. To identify appropriate construction of houses and building for energy saving and comfort.
6. To introduce Legislation for clean means of transport.
7. To reduce CO_2 emissions using RE in everyday life.

Within the program a Panel Discussion was held concerning Electricity, Energy Efficiency and Conservation, Building Comfort and Construction and Water situation and its rationalization chaired by Prof. Ali Sayigh, the speakers were:

1. **Dr. Adnan Mohammed Fakhro:** Deputy Chief Executive for Distribution & Customer Services.
2. **Mr. Jassim Isa Al Shirawi:** General Manager, Oil & Gas Affairs, **National Oil & Gas Authority (NOGA)**.
3. **Dr. Fuad Al-Ansari:** Dean of College of Engineering, University of Bahrain.
4. **Prof. Nader Al-Bastaki:** Chemical Engineering Dept., College of Engineering, University of Bahrain.

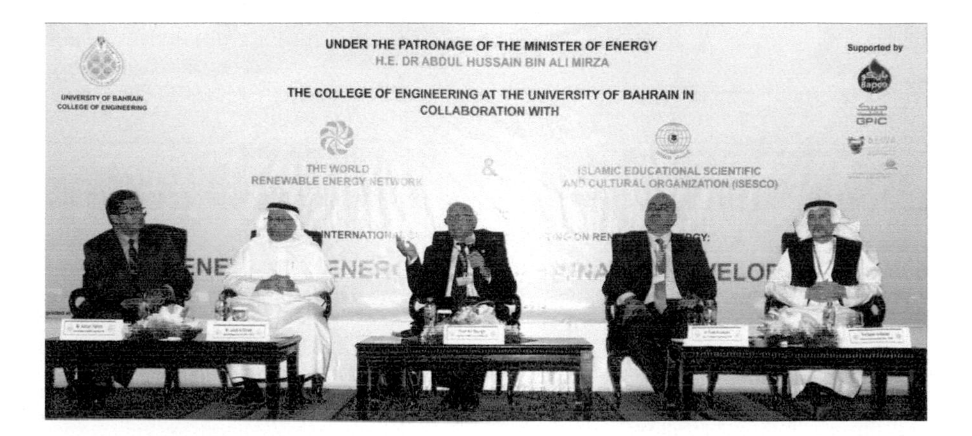

Keynote Speakers:

- Prof. Ali Sayigh: **Renewable energy is a reality in combating climate change**
- Dr. Lawrence Kazmerski: **Photovoltaic development and history**

Prof. Sayigh **Dr. Kazmerski**

Day 2:

- Prof. Waheed Al-Naser: **Renewable energy in the GCC**
- Prof. Marco Sala: 1. **Examples from Vernacular Architecture**
  2. **Building Design Modern Approach**
- Dr. Shaker Haji: **Fuel cells and their applications**
- Prof. Ali Sayigh: **Bahrain Feed-in-Tariff**

Dr. Haji **Prof. Sala**

Prof. Al-Naser **Prof. Sayigh**

Day 3:

- Mr. Adnan Naji: **BABCO & Bahrain PV Project – 5 MW**
- Prof. Issa Batarseh: **Solar Projects in the US**
- Dr. Khalid Hamid Bu-Rashid: **Electricity & Water in Bahrain**
- Dr. Gouri Datta: **Photovoltaic applications in India**
- Dr. Lawrence Kazmerski: **Progress on photovoltaic technology**

Dr. Datta

Dr. Kazmerski

Dr. Bu- Rashid

Mr. Naji

Prof. Batarseh

Day 4:

Three field visits were made: (i) to Bapco PV installations, (ii) the University PV Farm, (iii) GPIC Gulf Petrochemical Industries Co., (iv) Lunch and Distribution of the certificates at GPIC Club.

Recommendations

- That the economic, environmental, and social importance of renewable energy to all countries of the region be stressed and encouraged.
- That the appropriate legislation be adopted to facilitate the immediate utilization of all the abundant forms of renewable energy in the region.
- To adopt robust educational policies to ensure that renewable energy technologies are taught from high school to post-graduate levels.
- To enact a strong public relation policy to encourage commercial and private support for renewable energy.
- To facilitate a continuing series of conferences, seminars and meetings to ensure the dissemination of the most up-to-date technological advances in all fields of renewable energy.

9th ISESCO – WREN Agreement

During the period 23–25 March 2015, the 9th agreement between ISESCO and WREN for the period 2016–2017 was signed between Prof. Ali Sayigh and H E Deputy Director of ISESCO Dr. Amina Al-Hajri.

Preparatory mission of WREN to Indonesia April 2015

During the period April 5–14 a visit was made to Indonesia as part of the planning World Renewable Energy Congress – 15. Several meetings took place in Jakarta with the government officials responsible for renewable energy. Two lectures were delivered by Prof Sayigh, one to the Ministry of Energy officials and the other to university students in Bandung City.

Ministry of energy lecture

Prof. Sayigh and the minister of Energy with the Indonesian Organizing Committee

Visit to the UAE

During Prof Sayigh's stay in Sharjah, UAE, he visited the University of Sharjah and met with Prof Hamid M Al Naimiy.

Chancellor of University of Sharjah and President of Arab Union for Astronomy & Space Sciences. Also met the Dean of the College of Engineering Prof Sabah Alkass in presence of Prof Riadh Al-Dabbagh. They planned a World Renewable Energy Congress – 18, to take place in November 2017 at the University with the coordination of Dr Abdul Hai Al-Alami – College of Engineering, Sharjah University.

World Renewable Energy Congress – XIV
8–12 June 2015, Bucharest, Romania

Prior to the Congress on Sunday afternoon WREN Council took place at the Intercontinental Hotel from 3–5 p.m. Prof Sayigh briefed the Council about WREN activities during the last year. Also the Romanian Congress program was discussed and the Technical Committee was briefed about the speakers.

On 8 August, the Congress opening ceremony took place at the House of Parliament, Bucharest, in the presence of 350 participants. The program consisted of plenary speakers throughout the day, while the first part of the program 9:30–10:45 was devoted to the opening session. The speakers at the opening were:

- ***Prof. Ecaterina Andronescu*** – President of University Politehnica of Bucharest, Honorary Chair of WREC 2015
- ***Acad. Prof. Ionel Valentin Vlad*** – President of Romanian Academy, Honorary Chair of WREC 2015
- ***Prof. Ali Sayigh*** – President of WREN, Executive Chair of WREC 2015
- ***Assoc. Prof. Ruxandra Crutescu*** – Executive Co-Chair of WREC 2015
- ***Dr. Abdulaziz bin Othman Altwaijri*** – Director General of ISESCO
- ***Prof. Sorin Mihai Cîmpeanu*** – Minister of National Education

- ***Andrei Gerea*** – Minister of Energy and SME
- ***Dr. Tudor Constantinescu*** – Principal Adviser to the Director General for Energy, European Commission
- ***Mihnea Costoiu*** – Rector of University Politehnica of Bucharest
- ***Prof. Anton Anton*** – President of the Senate of Technical University of Civil Engineering of Bucharest
- ***Dr. Iulian Iancu*** – President of Industry Commission of the Chamber of Deputies, Romanian Parliament
- ***Dr. Mihnea Constantinescu*** – Special representative for Energy Security of the Romanian Government
- ***Prof. Gigel Paraschiv*** – Secretary of State, Ministry of Education and Scientific Research
- ***Prof. Tudor Prisecaru*** – President of ANCS (National Agency for Scientific Research)

Figure 1: Some technical Committee members at WREN Council meeting.

During the week, the program was divided into three parts. Part 1 was devoted to the plenary speakers from 9–10.30, followed by coffee break.

Figure 2: H E Dr. Altwaijri was welcomed by the President of the Senate of the University Polytechnic of Bucharest, the President of the University and the Deputy President.

Figure 3: Plenary speakers.

Prof. Ali Sayigh (UK),
Director General of WREN
and Congress Chairman

Dr. Akiba Segal (Israel),
Weizmann Institute
of Science

Dave Renné (USA),
President of International Solar
Energy Society (ISES)

Dr. Jinlong Gong (China),
Tianjin University

Prof. Horia Hangan (Canada),
The University of Western Ontario

Dr. Manuel Romero (Spain),
Deputy Director of Madrid Institute for
Advanced Research IMDEA Energy,
Vice-President of the International
Solar Energy Society and
President of the Spanish
Association of Solar Energy

Dr. Arch. Assoc. Prof. Ruxandra Crutescu (Romania),
Head of the Research-Development-Innovation
Department at Passivhaus EcoArchitect

Rainer Hinrichs-Rahlwes (EU),
Vice-President of European Renewable
Energies Federation

Prof. Carlos Guedes Soares (Portugal),
Universidade de Lisboa

Prof. Nataliia V. Voevodina (Russia),
Head of Scientific Dept. of
Far East Federal University

Prof. Ioan Stefănescu (Romania),
General Manager of the National Research &
Development Institute for Cryogenics and Isotopic
Technologies – ICIT Rm. Valcea

Dr. Larry Kazmerski (USA),
Director, Photovoltaics Centre,
National Renewable Energy Laboratory (NREL)

Dr. Soteris Kalogirou (Cyprus), Editor-in-Chief, Renewable Energy Journal Deputy Editor-in-Chief, Energy Journal

Dr. James Connolly (Spain), University of Valencia

Prof. Elias Stefanakos (USA), Director of the Clean Energy Research Center at the University of South Florida

Prof. Marco Sala (Italy), Florence University

On Tuesday afternoon 60 top speakers travelled to the Romanian Academy and were welcomed by the President and a lecture session was held. Among the speakers were Prof. Anne Grete Hestnes from Norway, and Dr. Larry Kazmerski.

Figure 4: Group photograph at the Romanian Academy.

Prof. Anne Grete Hestnes from Norway

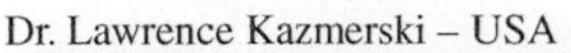
Dr. Lawrence Kazmerski – USA

Figure 5: The two main speakers at Romanian Academy.

The Technical Program of the Congress covered 10-topics and ran on Tuesday, Wednesday, Thursday, and Friday morning.

Topics

1. Photovolatic and Thermal Solar Technologies and Systems
2. Sustainable and Low Energy Architecture
3. Biomass, Biofuels, and Waste-to-Energy
4. Wind and Hybrid Energy

5. Water and Hydropower
6. Geothermal Energy
7. Hydrogen and Fuel Cells
8. Renewable System Integration
9. Policy, Finance, and Education
10. Energy Meteorology

On Friday 12 June the closing ceremony took place at the university. It was noted that the Congress had been attended by 300 participants from 40 countries, and over 220 abstracts had been received.

Figure 6 Some of the international speakers at the various technical sessions.

Ms. Diana Tutica – Romania

Ms. Irene R. Serrano – Spain

Dr. –Ing. Marc Roger – Germany

Ms. Salavatian – Iran, Prof. Sayigh, Prof. Aboulnaga – Egypt

Prof. Ionel-Valentin VLAD, President of and Romanian Academy, Dr. Sanda Carmen

Some of the Technical Sessions Participants

Participants from Mexico, Brazil, Turkey, and Portugal enjoying the Gala Dinner.

Prof. Emil Barbu Popescu President of Romanian University of Architecture

Prof. Viorica Curea – President of Romanian Union of Architects

Awards and Recognition

9 WREN awards were presented during the WREN Dinner on the first night of the Congress to:

H E Dr. Abdulaziz Bin Othman Altwaijri Director General of ISESCO, Rabat, Morocco. (H E also received a golden medal and a Diploma from the Senate President of University Polytechnic of Bucharest); Prof. Emil Popescu, Prof. Mihnea Costou, Prof. George Darie, Prof. Claudia Popescu, Prof. Fara,

Prof. Viorel Badescu, Dr. Ruxandra Crutescu, and Prof. Ecaterina Andronescu for their contributions toward a cleaner environment and the Congress. The WREN Dinner was co-sponsored by River Publishers and their Manager Mr. Mark de Jongh who was present at the Dinner and the Congress.
Prof. Sayigh presenting H E with a special certificate

Final Congress Recommendations: The Congress was successful in emphasizing to the appropriate Romanian authorities the urgency of adopting and implementing Renewable Energy. It was recommend that an annual symposium on Green Buildings and Renewable Energy in Romania should be held in April of each year under the name of WREN.

Organizers of the April 2016 Symposium in Bucharest

On 13 June a meeting took place between: Prof. Viorica Curea, President of Romanian Union of Architects, Prof. Ana-Maria Dabiji, Vice Rector of University of Architects, Prof. Manuela Epure, Vice Rector of Spiru Haret University, Dr. Ruxandra Crutescu of Spiru Haret University and Prof. Ali Sayigh to discuss the forthcoming symposium and a draft program was agreed.

MED GREEN FORUM – III (MGF)
Mediterranean Green Buildings and Renewable Energy, 26–28 August 2015, Florence, Italy

This World Renewable Energy (WREC) activity commenced in 2012 and will repeat itself every two years in one of the Mediterranean Countries. The initial MGF took place in Marseille, the second in Fez while this third one was held in Florence, and the MGF committee decided that MGF–IV July 2017 will once again be held in Florence. Future possible venues are Portugal, Malta and Greece.

Mission Statement: This is a World Renewable Energy Congress and Network Forum aimed at the international community as well as Mediterranean countries.

Each Forum highlights the importance of growing renewable energy applications in two main sectors: Electricity Generation and the Sustainable. The Mediterranean region was chosen to illustrate the viability of using renewable energy to satisfy all energy needs. We are hoping to demonstrate the effectiveness of using renewable energy in these countries to act as a beacon of light for the rest of the world to follow.

Renewables are the cornerstone and the foundation of a truly **sustainable energy future**. Our mission is to promote enabling policies and to further develop a broad range of renewable energy technologies and applications in all sectors – for electricity production, heating and cooling, agricultural applications, water desalination, industrial applications and for the transport sectors leading to A BETTER, CLEANER AND SAFER WORLD.

The 2015 Forum was hosted by the University of Florence – Department of Architecture represented by Professor Marco Sala and Professor Fernando Recalde (ABITA) as one of the partners, the second partner being Ms. Angela Grassi representing ETA and her two associates: Ms. Maddelena Grassi and Ms. Sabrina Palloni.

158 abstracts were received from around the world and there were 160 presentations from 53 countries. The Proceedings includes 82 full presented papers peer reviewed by the Technical Committee for their suitability for publication and will be published by Springer.

The Forum was opened by Prof. Ali Sayigh on 26 August 2015 explaining the motives and reasons behind the Forum and why it was being held in the region. Ms Aicha Bammoun – ISESCO representative, addressed the Forum on behalf of H E Dr. Abdulaziz Bin Othman Altwaijri – Director General of ISESCO one of the main sponsors of the Forum, outlining the importance of sustainability and clean environment as well as the strong bond between ISESCO and WREN in holding meetings around the world and explaining the goals of ISESCO in education, culture and science and environment. The opening ceremony was closed by Prof Marco Sala – Co-chairman of the Forum, expressing his delight in hosting the Forum and hoping everyone would derive great benefit from it by networking and learning from each other. Figure 1 shows the opening ceremoney and some of the participants in the main theatre.

Figure 1: Opening Ceremony.

This was followed by two presentations from Mr. Rainer Hinrichs-Rahlwes – (EREF), German Renewable Energy Federation (BEE), Germany, and Prof. Rajat Gupta, School of Architecture, Oxford Brookes University, UK, Figure 2.

Figure 2

In the second session there were four speakers: Mr. Bill Watts from Max Fordham, London; Prof. Marco Sala, Director of ABITA, Florence; Prof. Alessandra Battisti, PDTA Rome University and Prof. Wim Zeiler, TU Eindhoven, Netherlands (Figure 3).

Figure 3

Figure 3

Many participants actively participated asking questions and exchanging thoughts and ideas with the speakers during the day (Figure 4).

Figure 4

In the parallel session four top speakers presented their papers: Dr. David Renne, chairman of International Solar Energy Society, ISES, Boulder, USA; Prof. Dr. Joachim Pasel, Institute of Energy and Climate Research, Julich, Germany; Prof. Riadh Al-Dabbagh, Ajman University of Science and Technology, UAE and Dr. Ian Masters, Marine Energy Research Group, Swansea University, Wales, UK, see Figure 5.
Figure 5:

Dr. David Renne

Prof. Joachim Pasel

Prof. Riad Al-Dabbagh

Dr. Ian Masters

Figure 6: Shows some participants in Session 2.

Figure 7: A group photograph: *Back row*: Dr. Linda Hassaine, Algeria; Dr. Aicha Bammoun, ISESCO – Morocco; Prof. Anwar El Hadi, Sudan; Prof. Ali Sayigh – Forum Chairman, UK; Dr. Salam Darwish, UK; Dr. Kamil Yousif, Iraq; Dr. Fuad Al-Ansari, Bahrain; Dr. Herliyani Suharta, Indonesia. *Front*

Figure 7

row: Dr. Shadia Ikhmayis, Jordan; Prof. Riad Al-Dabbagh, Ajman, UAE; Prof. Najma Laaroussi, Morocco; Dr. Mounir Ouzguenda, Algeria, Dr. Shakir Sh. Ali, Dubai, Germany; Prof. Mohsen Aboulnaga, Egypt and Prof. Marco Sala – Forum Co-chair.

Women scientist have always played an important role in all WREC/WREN activities and many participated in this Forum, see following figures. (Figure 8)

Figure 8: Some women participants at the Forum with Prof. Yoshinori Itaya.

During the Forum there was a WREN-Springer sponsored dinner for 60 invited speakers and VIPs held at Prof Sala's residence in the hills of Florence. MED Green Forum is grateful for Springer in Sponsoring with WREN the dinner, thanks in particular to Ms. Tiffany Gasbarrini, Senior Editor, Energy & Power. Springer also had a book exhibit at the Forum (Figure 9).

Figure 9

Below are some the presenters from the Thursday sessions.

Dr. Herman Darnel Ibrahim
Host Chairman of WREC – 15, Indonesia (Figure 10)

Dr. Eng. Angelo Freni – Italy

Prof. Myrsini Christou – Greece

Dr. Neveen Hamza – UK

Prof. Manuel Guedes – Portugal

Prof. A. K. M. Sadrul Islam – Bangladesh

One of the highlights of the Forum was to invite students from many parts of Europe and the Mediterranean Region to present their architectural projects.

Figure 11

Figure 11 Students Posters and Models with some of the participants, among them Dr. Kuruvilla Mathew, host Chairman of WREC – Australia, Dr. Aicha Bammoun ISESCO representative and Dr. Herliyani Suharta Co-host Chair of Indonesian WREC – 15.

On Thursday evening, the Med Green Forum – III, Gala Dinner was held at a restaurant overlooking Florence. (Figure 12)

Figure 12: Pictures from the Gala Dinner.

On Friday, the last day of the MGF-III, there were two parallel sessions, one mostly devoted to Italian building industries while the other was devoted to the participant presentations. Several papers were presented by young architects and scientists.

Figure 13: Speakers at the Forum.

A post-Forum industrial trip was organized by Prof. Fernando which was considered a great success by all those who took part.

At the closing ceremony Prof. Sayigh with Prof. Marco Sala presented the Students awards and hoped the participants had found the Forum beneficial. My thanks goes to all the participants for their efforts in making this Forum a very rewarding and enjoyable meeting. Thanks to the Technical Committee and their chair: Prof. Fernando Recalde; Mr. Rainer Hinrichs-Rahlwes; Dr Neveen Hamza; Prof Bahram Moshfegh; Prof Fernando Butera; Prof. Run-ming Yao and Dr. Fuad Al-Ansari and Co-chair Prof. Marco Sala; all sessions' chairs; and special thanks to ETA and Ms. Angela Grassi for doing such an admirable job of all the administration. My thanks to the Italian Industry representatives and to Prof. Recalde for organizing their talks and the visit on Saturday.

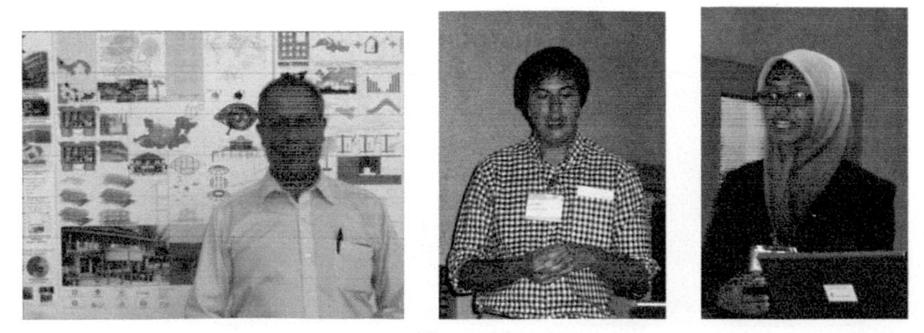

Figure 14

A special thanks to all the Sponsors: Italian Trade Agency; Italian Ministry of Foreign Affairs; Municipality of Florence; University of Florence; Springer; ISESCO; and WREN.

Prof. Mohsen Aboulnaga has prepared an excellent PDF report about the MGF- III Event which can be found on the internet:

http://culturaegitto.blogspot.com.eg/2015/08/med-green-forum-2015-mediterranean.html/.2015-mediterranean.html/

ISESCO Ministers of Environment Conference October 2015

The Conference was opened by Her H. Princess Hasina of Morocco. A two day discussion and papers were presented about the importance of renewable energy and cleaner environment not only in the ISESCO countries but also globally.

In the right hand side H. H. Princess Hasina of Morocco with Prof. Sayigh

Ministers of environment, ISESCO attending the conference

World Renewable Energy Congress/Network, WREC/WREN

International Seminar in Britain, No. 90, 8-14 November 2015, Old Ship Hotel Brighton, UK

FINAL REPORT

RENEWABLE ENERGY

Policy, Security, Electricity, Transport, Water Resources, and the Built Environment

Among the Sponsors:

WREC/WREN

River Publishers

Swift-Hook Associates

The Seminar participants were from fourteen different countries, UK, Spain, Romania, Finland, Poland, Philippine, China, Sweden, Iraq, Bahrain, UAE, Egypt, Iran, and Malaysia. It was sponsored by Islamic Educational Scientific and Cultural Organization (ISESCO), World Renewable Energy Congress and Network (WREC/WREN), River Publishers and Swift-Hook Associates. The seminar ran from 9–14 November 2015 at the Old Ship Hotel, Brighton, UK.

The Seminar was directed by Prof. Ali Sayigh, chairman of WREC and director general of WREN.

Participants arrived and registered on 8 November and a welcome dinner was given by WREN that evening see below:

On Monday Prof. Sayigh read the opening speech of H E Dr. Abdul Aziz Altwaijri – Director General of ISESCO and thanked HE for ISESCO support in giving the opportunity to attend the seminar to several participants and speakers. He also thanked the other sponsors for their generous support and explained the program to the participants.

More than 20 1-h presentations were given by different speakers throughout the week.

Monday was devoted to: **Sustainability & Renewable Energy Options-I**

The speakers were Prof. Phil Eames from Loughborough University, UK; Prof. Runming Yao, from Reading University and China; Dr. Ruxandra Crutescu from Spiru Haret University, Romania; Dr. Arthur Williams, from Nottingham University, UK; Mr. Bill Watts from Max Fordham, UK; Dr. Ala Hasan from VTT Technical Research Centre of Finland, Finland; and Dr. Jazaer Dawody from Volvo Group, Sweden;

Prof. Phil Eames

Prof. Runming Yaw

Dr. Ruxandra Crutescu

Dr. Arthur William

Mr. Bill Watts

Dr. Ala Hasan

Dr. Jazaer Dawody

Participants and lecturers at a session

On Tuesday, the theme of the Seminar was: **Renewable Energy Options-II**

Among the speakers were Prof. Dorota Chwieduk, Warsaw University, Poland; Mr. David Milborrow, Lewes, UK; Prof. Robert Critoph, University of Warwick, UK; Prof. Giuliano Premier, University of Glamorgan, Wales, UK; Dr. A Kaabi Nejadian, Tehran, Iran, and Prof. Don Swift-Hook, University of Kingston, UK.

Prof. Dorota Chwieduk

Mr. David Milborrow

Prof. Robert Critoph

Prof. Giuliano Premier Dr. A Kaabi Nejadian Prof. Don Swift-Hook

On Wednesday, a cultural and scientific trip was made to London after having lunch at the home of Prof. Swift-Hook where 3-kW of PV installed 2 years ago.

In London, we looked at Kings Cross Station where 250 kW of PV was installed 3-years ago. See below:

Then we went to the British Museum for 2.5 h visit, see below

On Thursday the theme was: **National and International ProgramS**, and the speakers were : Prof. Riadh Al-Dabbagh, Ajman University, UAE; Mr. David Milborrow second lecture; Prof. Nader Al-Bastaki, Bahrain University, Bahrain; Dr. Vahid Tabatabai from Tomorrow's Energy Ltd, Cardiff, Wales, UK; and Prof. David Elliott from Open University, Milton Keen, UK.

Prof. R Al-Dabbagh Prof. Nader Al-Bastaki Dr. Vahid Tabatabai Prof. David Elliott

On Friday the theme of the Seminar was: **Renewable Energy Options-III**

Among the speakers were Mr. Nicholas Dunlop, from Climate Parliament, Brighton, UK; Mr. Tony Book, ex-Direrctor of Riomay, Brighton, UK; Dr. Salam Darwish, Phoenix Renewable Energy Centre, Manchester, UK;

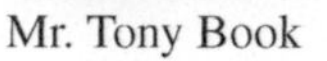

Mr. Nicholas Dunlop | Mr. Tony Book | Dr. Salam Darwish | Dr. Ian Masters

Dr. Ian Masters, Swansea University, Wales, UK and Prof. Don Swift-Hook, University of Kingston, UK.

A dinner was held on Friday evening when all participants received their certificate of attendance, a USB containing all the lectures and a group photograph.

World Renewable Energy Congress – XVII, Bahrain 2016

In late November Prof Sayigh made a trip to Bahrain to finalize the World Renewable Energy Congress – 17, 3–8 December 2016, with the University of Bahrain as a partner and host organization.

See below Prof. Sayigh with the Ex-President of Bahrain University Dr. Ibrahim Al-Jannahi.

The main organizers of the Congress: Prof. Waheeb Alnaser – Deputy President of Bahrain University for Scientific Affairs; Dr. Fuad Alansari – Dean of College of Engineering, University of Bahrain; and Prof. Nader Al-Bastaki, Department of Chemical Engineering, Bahrain University.

World Renewable Energy Congress & WREN Calendar of Events for 2016

- **2–3 March: Ajman Fourth Environment Conference**

  Ajman University, Ajman, UAE
  Contact: khaledmhod565@hotmail.com

- **10–12 April: European Seminar for Sustainable Architecture**

  Intercontinental Hotel, Bucharest, Romania.
  Organized by: Academy of Technical sciences of Romania, The Union of Architects of Romania
  World Renewable Energy Network and Spiru Haret University
  Contact: Prof. Manuela Epure, prorector_cercetare@spiruharet.ro

- **24–30 July: WREN-ISESCO International Seminar in Britain: No. 92**

  Renewable Energy Policy, Security, Electricity, Transport, Water Resources, and the Built Environment
  Old Ship Hotel, Brighton, UK.
  Contact: Prof. Ali Sayigh, asayigh@wrenuk.co.uk

- **19–23 September: World Renewable Energy Congress – XV, Sustainable Energy for All and All for Sustainable Energy.**

  Jakarta Convention Center, Jakarta, Indonesia.
  Contact: www.wrec2016indonesia.com

- **3–8 December: World Renewable Energy Congress – XVII, Renewable Energy, Sustainability, and Clean Environment.**

  Crowne Plaza Hotel, Bahrain
  Contact: Dr. Fuad Al-Ansari, Dean of Engineering College, falansari@uob.edu.bh

8

Calendar of Events WREN: 2016 and 2017

2016

2–3 March
Ajman Fourth Environment Conference: Ajman University, Ajman, UAE Eng
Contact: khaledmhod565@hotmail.com

11–13 April
European Seminar for Sustainable Architecture: Intercontinental Hotel, Bucharest, Romania
Organized by Academy of Technical Sciences of Romania, The Union of Architects of Romania; World Renewable Energy Network, and
Spiru Haret University
Contact: Prof. Manuela Epure, prorector_cercetare@spiruharet.ro

23–29 July
WREN-ISESCO International Seminar in Britain: No. 92, Renewable Energy Policy, Security, Electricity, Transport, Water Resources and the Built Environment, Old Ship Hotel, Brighton, UK
Contact: Prof. Ali Sayigh, asayigh@wrenuk.co.uk

19–23 September
World Renewable Energy Congress – XV, Sustainable Energy
for All and All for Sustainable Energy, Jakarta Convention Center, Jakarta, Indonesia
Contact: www.wrec2016indonesia.com

3–8 December
World Renewable Energy Congress – XVII, Renewable Energy,
Sustainable Buildings and Clean Environment, Crowne Plaza Hotel, Bahrain
Contact: Dr. Fuad Al-Ansari, Dean of Engineering College,
falansari@uob.edu.bh

2017

5–8 February
World Renewable Energy Congress – XVI, Perth, Australia, in cooperation with Murdoch University, Perth, Australia
Contact: Dr. Kuruvilla Mathew, School of Engineering and Information Technology, Murdoch University, Murdoch, WA 6150, USA
k.mathew@murdoch.edu.au

29 July–2 August 2017
Mediterranean Green Forum – IV, MGF-4, School of Architecture, Florence, Italy
Contact: Prof. Ali Sayigh, asayigh@wrenuk.co.uk

10–16 September 2017
WREN-ISESCO International Seminar in Britain, Brighton, UK
Contact: Prof. Ali Sayigh, asayigh@wrenuk.co.uk

11–16 November 2017
World Renewable Energy Congress – XVIII, University of Sharjah, UAE
Contact: Dr. Abdul Hai Al-Alami, aalalami@sharjah.ac.ae

For more details go to www.wrenuk.co.uk

Index

About the Editor

Ali Sayigh, a British Citizen, Graduated from London University and Imperial College, and acquired following degrees: B.SC, DIC, Ph.D., AWP, and finally CEng in 1966. He is holding various prestigious positions such as Fellow of the Institute of Energy and the Institution of Engineering and Technology (previously called IEE), Chartered Engineer, and Chairman of Iraq Energy Institute.

Prof. Sayigh was working at Baghdad University, College of Engineering, King Saud University, College of Engineering, Saudi Arabia as a full-time Professor and also worked in Kuwait University as a part-time professor. He is also the Head of Energy Department at Kuwait Institute for Scientific Research (KISR) and Expert in renewable energy at AOPEC.

He started working in solar energy since September 1969. In 1984, he published with Pergamon Press his first International Journal for Solar and Wind Technology as an Editor-in-Chief. From 1990 up to April 2014, he was the Editor-in-Chief of Renewable Energy Journal incorporating Solar and Wind Technology, published by Elsevier Science Ltd, Oxford, UK. He is also the Editor-in-chief of comprehensive renewable energy, 8 volumes with 154 contributors.

He is the Founder and Chairman of the ARAB Section of ISES since 1979, was chairman of UK Solar Energy Society for 3-years and consultants to many national and international organizations, among them UNESCO, ISESCO, UNDP, ESCWA, and UNIDO.

Since 1977, Prof. Sayigh organized and directed several Renewable Energy Conferences and Workshops on ICTP – Trieste, Italy, Canada, Colombia, Algeria, Kuwait, Bahrain, Malaysia, Zambia, Malawi, India, West Indies, Tunisia, Indonesia, Libya, Taiwan, UAE, Oman, Czech Republic, West Indies, Bahrain, Germany, Australia, Poland, Netherlands, Thailand, Oman, Korea, Iran, Syria, Saudi Arabia, Singapore, USA, and UK.

In 1990, he established the World Renewable Energy Congress and in 1992 the Network (WREN) which have their Congresses every 2 years, attracting more than 100 countries each time. In 2000, he and others in UAE, Sharjah founded Arab Science and Technology Foundation (ASTF). He is now the Chairman of Iraq Energy Institute.

He edited, contributed, and written more than 35 books and more than 600-papers in various international journals and Conferences. Winner of PROSE award in 2012 for the best book entitled Comprehensive Renewable Energy, which involved 154 contributors from 80 countries. His recent book came September 2013, Sustainability, Energy and Architecture published by Elsevier. He established Med Green Forum (Mediterranean Green Buildings and Renewable Energy Forum in 201.